Health & Wellness

GRADE 3

Linda Meeks
The Ohio State University

Philip Heit
The Ohio State University

McGraw Hill Education

Front Cover: McGraw-Hill Education

Title Page: McGraw-Hill Education

Back Cover: McGraw-Hill Education

MHEonline.com

Send inquiries to:
McGraw-Hill Education
8787 Orion Place
Columbus, OH 43240

ISBN: 978-0-07-671448-3
MHID: 0-07-671448-9

Printed in the United States of America.

13 LWI 25 24 23

Contributors

Celan Alo, M.D., MPH
Medical Epidemiologist
Bureau of Chronic Disease and Tobacco Prevention
Texas Department of Health
Austin, Texas

Danny Ballard, Ed.D.
Associate Professor, Health
Texas A&M University
College of Education
College Station, Texas

Lucille Villegas Barrera, M.E.D.
Elementary Science Specialist
Houston Independent School District
Houston, Texas

Gus T. Dalis, Ed.D.
Consultant of Health Education
Torrance, California

Alisa Evans-Debnam, MPH
Dean of Health Programs
Fayetteville Technical Community College
Fayetteville, North Carolina

Susan C. Giarratano-Russell, MSPH, Ed.D., CHES
Health Education, Evaluation & Media Consultant
National Center for Chronic Disease Prevention & Health Promotion
Centers for Disease Control & Prevention
Glendale, California

Donna Lloyd-Kolkin, Ph.D.
Principal Associate
Public Health Applications & Research
Abt Associates, Inc.
Bethesda, Maryland

Mulugheta Teferi, M.A.
Principal
Gateway Middle School
Center for Math, Science & Technology
St. Louis, Missouri

Roberto P. Treviño, M.D.
Director, Social & Health Research Center
Bienestar School-Based Diabetes Prevention Program
San Antonio, Texas

Dinah Zike, M.Ed.
Dinah Might Adventures LP
San Antonio, Texas

Content Reviewers

Mark Anderson
Supervisor, Health Physical Education
Cobb County Public Schools
Marietta, Georgia

Ken Ascoli
Assistant Principal
Our Lady of Fatima High School
Warren, Rhode Island

Jane Beougher, Ph.D.
Professor Emeritus of Health Education, Physical Education, and Education
Capital University
Westerville, Ohio

Lillie Burns
HIV/AIDS Prevention Education
Education Program Coordinator
Louisiana Department of Education
Baton Rouge, Louisiana

Jill English, Ph.D., CHES
Professor, Soka University
Aliso Viejo, California

Elizabeth Gallun, M.A.
Specialist, Comprehensive Health Education
Maryland State Department of Education
Baltimore, Maryland

Brenda Garza
Health Communications Specialist
Centers for Disease Control and Prevention
Atlanta, Georgia

Sheryl Gotts, M.S.
Consultant, Retired from Milwaukee Schools
Milwaukee, Wisconsin

Russell Henke, M.Ed.
Coordinator of Health
Montgomery County Public Schools
Rockville, Maryland

Kathy Kent
Health and Physical Education Teacher
Simpsonville Elementary School at Morton Place
Simpsonville, South Carolina

Bill Moser, M.S.
Program Specialist for Health and Character Education
Winston-Salem Forsyth City Schools
Winston-Salem, North Carolina

Debra Ogden
Curriculum Coordinator
District School Board of Collier County
Naples, Florida

Thurman Robins
Chair/Professor
Health and Kinesiology Department
Texas Southern University
Houston, Texas

Sherman Sowby, Ph.D., CHES
Professor, Department of Health Science
California State University, Fresno
Fresno, California

Greg Stockton
Health and Safety Expert
American Red Cross
Washington, D.C.

Deitra Wengert, Ph.D., CHES
Professor, Department of Health Science
Towson University
Towson, Maryland

Susan Wooley-Goekler, Ph.D., CHES
Adjunct Faculty
Kent State University
Kent, Ohio

Medical Reviewers

Celan Alo, M.D., MPH
Medical Epidemiologist
Bureau of Chronic Disease and Tobacco Prevention
Texas Department of Health
Austin, Texas

Donna Bacchi, M.D., MPH
Associate Professor of Pediatrics
Director, Division of Community Pediatrics
Texas Tech University Health Science Center
Lubbock, Texas

Olga Dominguez Satterwhite, R.D., L.D.
Registered Dietitian and Diabetes Educator
Baylor College of Medicine
Houston, Texas

Roberto P. Treviño, M.D.
Director, Social & Health Research Center
Bienestar School-Based Diabetes Prevention Program
San Antonio, Texas

Contents

UNIT A Mental, Emotional, Family, and Social Health

CHAPTER 1 Mental and Emotional Health

CHAPTER 2 Family and Social Health

UNIT B Growth and Nutrition

CHAPTER 3 Growth and Development

CHAPTER 4 Nutrition

UNIT C Personal Health and Safety

CHAPTER 5 Personal Health and Physical Activity

CHAPTER 6 Violence and Injury Prevention

UNIT D

Drugs and Disease Prevention

CHAPTER 7 Alcohol, Tobacco, and Other Drugs

CHAPTER 8 Communicable and Chronic Diseases

UNIT E Community and Environmental Health

CHAPTER 9 Consumer and Community Health

CHAPTER 10 Environmental Health

Features and Activities

UNIT A Mental, Emotional, Family, and Social Health

ACTIVITIES

UNIT B Growth and Nutrition

ACTIVITIES

ACTIVITIES

UNIT C Personal Health and Safety

UNIT D Drugs and Disease Prevention

Features and Activities

UNIT E Community and Environmental Health

Life Skills

Life Skills are actions you can take to improve and maintain your health. The life skills that are taught in this text are listed below.

Using Foldables™

To help you learn these life skills, each of the Learning Life Skills features in this book include Foldables™. Foldables™ are three-dimensional graphic organizers you will make. They will help you understand the main points of each life skill.

UNIT A

Mental, Emotional, Family, and Social Health

CHAPTER 1

Mental and Emotional Health

What Do You Know?

What do you know about mental and emotional health? Read the statements. Answer each one **yes** or **no**.

___?___ Talking about your feelings is healthful.

___?___ Your personality includes how you look, think, act, and feel.

___?___ Riding your bicycle without a helmet is a risk behavior.

___?___ People with good character don't mind if they break rules.

The answer to the first three statements is **yes**. The answer to the fourth statement is **no**. How did you do? You will learn more as you read this chapter, **Mental and Emotional Health**.

LESSON 1

Health Is Well-Being

You will learn . . .

- about the three parts of health.
- how choosing healthful behaviors protects your health.
- how choosing risk behaviors can harm your health.
- ten life skills that can help you take responsibility for your health.

Vocabulary

- **health**, *A5*
- **life skill**, *A7*
- **health goal**, *A8*
- **health behavior contract**, *A8*

What is well-being? It is having a healthy body. It is having a healthy mind. It is getting along with others.

The Three Parts of Health

Health is more than being free from illness. Health has three parts. **Health** is the condition of your body, mind, and relationships. All three parts of health affect your well-being.

The way you act, or *behave*, affects the three parts of health. You can learn how to make wise health choices. You can even set goals for better health. Behaving in healthful ways helps your health. This shows that you are responsible for your health.

What are the three parts of health?

Your well-being depends on the three parts of health working together.

Mental and emotional health is how well your mind works. It is how well you express your emotions. It is also how you feel about yourself.

Family and social health is how well you get along with others. It involves your relationships with family members, friends, and people in the community.

Physical health is how well your body works. A healthy body helps you do the things you want to do.

Healthful Behavior

Healthful behaviors are actions that can improve or protect your health. *Risk behaviors* are actions that can harm your health. When you brush your teeth, you practice a healthful behavior. You also take responsibility for your health. The chart below gives examples of both healthful and risk behaviors.

Wearing a bicycle helmet is a healthful behavior. In what other ways can you take responsibility for your health?

Compare

Part of Health	Healthful Behavior	Risk Behavior
Physical Health	• Get enough sleep • Eat healthful foods • Play outdoors	• Stay up late • Eat foods high in fat and sugar • Watch too much television
Mental and Emotional Health	• Talk about your feelings • Do activities you enjoy • Learn about new things	• Ignore your feelings • Be bored • Refuse to try new things
Family and Social Health	• Help family members • Be kind to your friends • Avoid arguments with others	• Complain when a family member asks for help • Talk about your friends behind their backs • Fight with others

Life Skills for Health

A **life skill** is a healthful action you learn and practice for life. Each life skill gives you a way to take responsibility for your health. This book can help you learn how to use the ten life skills listed below.

- Access Health Facts, Products, and Services
- Practice Healthful Behaviors
- Manage Stress
- Analyze What Influences Your Health
- Use Communication Skills
- Use Resistance Skills
- Resolve Conflicts
- Set Health Goals
- Make Responsible Decisions
- Be a Health Advocate

What is a life skill?

Write About It!

Take Responsibility
Write a paragraph describing one way to take responsibility for health. Explain how this action can help you improve your physical, mental and emotional, and family and social health. Share your paragraph with a classmate.

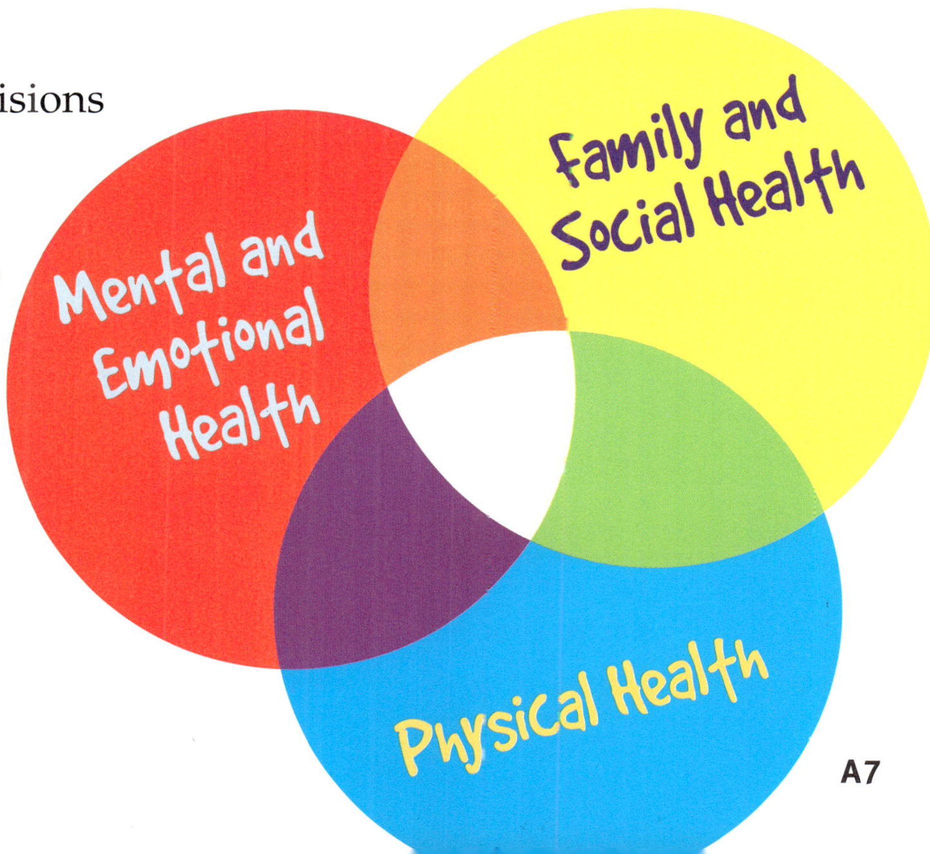

Set Health Goals

▼ A health behavior contract can help you set and meet goals. In step 3, tell who might help you reach your goal—a parent or guardian, a teacher, a family doctor, a school nurse, or a coach.

Setting health goals is an important part of health. A **health goal** is something you work toward in trying to stay healthy. You can use a health behavior contract to make a plan to reach a health goal. A **health behavior contract** is a written plan to help you practice a healthful behavior. Making a health behavior contract means taking responsibility for your health.

What is a health goal?

Health Behavior Contract

Name ____________________ Date ____________

1 **Health Goal** I will get plenty of physical activity.

2 **Effect on My Health** Physical activity helps keep my bones and muscles strong. Physical activity helps me stay at a healthful weight.

3 **My Plan** I will ride my bike for 30 minutes each day for a week. My parent or guardian can help me. Each day that I am physically active, I will draw a smiley face on my calendar.

My Calendar

Sun.	Mon.	Tues.	Wed.	Thurs.	Fri.	Sat.
☺		☺		☺		☺

4 **How My Plan Worked** I will tell if I was physically active each day. If I was not active, I will tell why. I will find ways to be active even in bad weather.

1. Identify the health goal you want to set.

2. Explain how your goal might affect your health.

3. Describe a plan you will follow to reach your goal. Keep track of your progress. Tell who can help you.

4. Evaluate how your plan worked. Decide what to do if your plan failed.

ACTIVITY

LIFE SKILLS

CRITICAL THINKING

Set Health Goals

You want to improve your physical health. Begin by making a health behavior contract like the one on page A8. Then follow the steps for setting health goals.

1. **Write the health goal you want to set.** I will get plenty of physical activity.
2. **Explain how your goal might affect your health.** In what ways can getting plenty of physical activity make you healthier?
3. **Describe a plan you will follow to reach your goal. Keep track of your progress.** Choose a physical activity you like. Plan to do it every day. Decide how you will keep track of your progress. Will you make a chart? Will you keep a log? Talk about your plan with a parent or guardian. Ask him or her to help you practice your health goal. Follow your plan for one week.
4. **Evaluate how your plan worked.** Are you following your health behavior contract? If not, what can you do to make your plan work?

LESSON REVIEW

Review Concepts

1. **Explain** the difference between mental and emotional health and family and social health.
2. **Explain** why riding a bicycle without a helmet is a risk behavior.
3. **Name** three healthful behaviors that you should do each day.

Critical Thinking

4. **Apply** Name three ways that you can take responsibility for your three parts of health.
5. LIFE SKILLS **Set Health Goals** Suppose you set a health goal to eat healthful snacks. What are four steps you can follow to reach this health goal?

LESSON 2

Your Self-Concept

You will learn . . .

- what makes you special.
- actions that help you have a healthful self-concept.
- about different ways that people learn.

Vocabulary

- **personality**, *A11*
- **heredity**, *A11*
- **self-concept**, *A12*
- **respect**, *A12*
- **rule**, *A12*

Think about the people you know. You are like some of those people in certain ways. In many ways, though, you are different. That's because there is only one you. You are one of a kind!

You Are Special

About 5 billion people live on planet Earth. Each one is special, including you! There never was and never will be another you. Your heredity, personality, talents, and skills make you special. No one else has exactly the same combination of these things as you.

What is heredity?

ACTIVITY

Art LINK

Self-Portrait

Draw a picture of yourself doing something you enjoy. Write a sentence telling why you enjoy this activity.

Your talents and skills make you special. You may have musical or artistic talent. You may have skill in math or in certain sports.

Your personality makes you special. Your **personality** includes how you look, think, act, and feel. You may be shy or outgoing. You may like some parts of your personality. Other parts you may want to change.

Your heredity makes you special. Your **heredity** (huh•RED•ih•tee) is the traits you get from your birth parents. The color of your eyes and your height are two of these traits.

A Healthful Self-Concept

Your **self-concept** is the feelings you have about yourself. When you have a healthful self-concept, you have respect for yourself. Here are some actions that show you have a healthful self-concept.

- **Show respect for yourself and others.** **Respect** is treating others as you want to be treated. *Self-respect* is thinking highly of yourself. When you respect others, you treat them fairly. Following rules is one way to show respect for others. A **rule** is a guide to help you do the right thing.

When you have self-respect, you recognize your strengths. You learn from your mistakes. You work on your weaknesses.

ACTIVITY

Music LINK

Respectful Rap

Write a rap song that describes how to show respect for others. Perform your rap for the class.

Design Pics/Ron Nickel

▶ **Taking care of your belongings shows self-respect.**

- **Practice life skills for health.** This shows that you care about your health. Respecting your health is part of respecting yourself.
- **Do your best.** You do some things very well. These are your strengths. You are not as good at other things. These are your weaknesses. Doing your best means working to improve both your strengths and weaknesses. This helps you to have a healthful self-concept.

Healthful Choices

Your actions show that you have self-respect and a healthful self-concept. You make responsible choices about how you live your life. You choose healthful behaviors over risk behaviors. You care about your health.

How does practicing life skills show self-respect?

ACTIVITY

Health Online

Feeling Good About You

People with a healthful self-concept make healthful choices. Research the relationship between self-concept and choice. Use the e-Journal writing tool to write a report on your findings.

▼ Grooming is an action that shows self-respect.

Practice Healthful Behavior

Setting goals and practicing healthful behaviors can help you do your best. To practice healthful behaviors, follow the steps listed on the clipboard.

Practice Healthful Behaviors

1. Learn about a healthful behavior.
2. Practice the behavior.
3. Ask for help if you need it.
4. Make the behavior a habit.

Learning Styles

Do you do your best to study and learn? Did you know that people study and learn in different ways? Some people learn best by listening. Some people learn best by doing what's required. Others learn best by writing things down. There are several different ways to learn. These are called *learning styles*. If you find your learning style, you can improve your grades.

Name one way to do your best.

Ask questions when you are unsure of something in class. This shows that you care about your schoolwork and want to do your best.

ACTIVITY

LIFE SKILLS **CRITICAL THINKING**

Practice Healthful Behaviors

Practicing healthful behaviors can help you learn. Use this activity to get started.

1. **Learn about a healthful behavior.** Make a list of activities that can help you learn. You might list studying with a friend or reading aloud. Choose one activity that seems to work well for you.
2. **Practice the behavior.** Practice the activity for three days. Keep a record to make sure you do the behavior each day.
3. **Ask for help if you need it. A parent or guardian can help.**
4. **Make the behavior a habit.** Focus on the activity you choose. Then pick a different way to learn and try again.

LESSON REVIEW

Review Concepts

1. **Name** what makes a person special.
2. **Identify** actions you can take to have a healthful self-concept.
3. **Define** learning style.

Critical Thinking

4. **Synthesize** You have tried to learn multiplication by reading. What other methods could you use to learn it?
5. LIFE SKILLS **Practice Healthful Behaviors** You want to learn the correct way to floss your teeth. Who could help you?

LESSON 3

Your Mind and Emotions

You will learn . . .

- how to share emotions in healthful ways.
- how to use I-messages.
- ways to stay in a good mood.
- ways to keep your mind healthy.

Vocabulary

- **emotion**, *A17*
- **self-control**, *A18*
- **I-message**, *A18*
- **attitude**, *A20*

Look at this picture. You can tell what the girl is doing. Can you tell what she is feeling? You can if you look closely at her face. The expression on her face shows how she feels.

Pixtal/Age Fotostock

Your Emotions

To maintain health, you need to show emotion in healthful ways. An **emotion** is a feeling inside you. There are many kinds of emotions. Here are some healthful ways to express emotions.

- **Fear** Talk to your parents or guardian. Be honest. The adult may be able to help you get over your fear.
- **Caring** Caring is showing concern for others. If you care for someone, tell him or her. He or she may feel the same way about you.
- **Sadness or Grief** *Grief* means a feeling of being very sad. People feel grief over a loss, such as when a pet dies. They may feel grief when they are disappointed. Let your emotions out by crying. Talk to a responsible adult about the pain you feel. Keep a private journal.
- **Joy and Happiness** When you feel joy and happiness, show it. Don't try to hide how you feel. Smile or laugh. Your joy and happiness could make others feel happy.

What are emotions?

ACTIVITY

BUILD Character

Show You Care

You show you care when you help someone. Think of a way to show a friend that you care. Make a picture collage of what you would do. Then carry out your plan.

Angry Feelings

Sending I-messages

- Tell what has happened.
- Tell how it affects you.
- Tell how you feel.

Suppose a friend makes you angry. You need to share your angry feelings in healthful ways.

Self-control helps you share your angry feelings in a healthful way. **Self-control** is having control over your emotions and actions. Taking deep breaths and counting to ten may help you keep your self-control when you are angry. Another way to express angry feelings is by using an I-message. An **I-message** is a healthful way to talk about feelings when you are upset.

What is self-control?

Suppose you are playing tennis. You get hit with the ball. You feel angry. You might use the I-message below.

What Happened	How It Affected You	How You Felt
When the tennis ball hit me . . .	. . . it hurt . . .	. . . and I felt angry.

How You Feel

Your *mood* is the way you feel at a certain time. There are ways to stay in a good mood.

- **Get physical activity.** Your brain makes special chemicals when you exercise. These can lift your mood.
- **Do something well.** Being successful makes you feel good about yourself.
- **Spend time with people who care about you.** These people can help you stay in a good mood during hard times.
- **Help someone else.** This helps you both stay in a good mood.

How can you stay in a good mood?

ACTIVITY LIFE SKILLS

CRITICAL THINKING

Use Communication Skills

Your brother used your baseball mitt without asking. You are angry. Use communication skills to share your emotions. Work with a partner to role-play these steps.

1. **Choose the best way to communicate.** How will you tell your brother you are angry? Will you yell and scream? Will you talk about how you are feeling?
2. **Send a clear message. Be polite.** Write an I-message to tell your brother how you feel. Be direct.
3. **Listen to each other.** Say your I-message. Then listen to your brother. Maybe he had a good reason for using your mitt.
4. **Make sure you understand each other.** Tell your brother what you heard. Ask what he heard from you. How can you improve your communication skills?

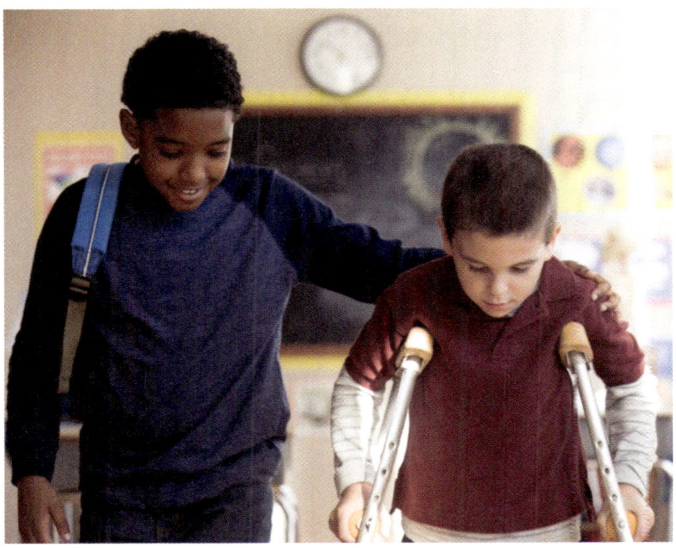

▶ Helping someone can help you both stay in a good mood.

A Healthy Mind

A healthful attitude helps keep your mind healthy. Your **attitude** is the way you think, act, or feel.

If you have a healthful attitude, you think about things in a healthful way. Suppose your parents or guardian tell you that you have to do chores. You do the chores as a way to help your family. You carry them out cheerfully. You have a healthful attitude. These are ways to keep your mind healthy.

ACTIVITY
Consumer Wise

Media Messages

Sometimes ads cause people to worry. Work with a partner. Cut out ads from old newspapers and magazines. Which ads cause you to worry? Which don't? Paste the ads on poster board to show how they are different.

Avoid Worry

You may worry before you take a test. You may feel anxious when you perform a play for a group of people. Worrying too much can harm your mental and emotional health. Worry also can harm your physical health. It can cause your body to feel tense and tired. It may cause you to become ill. You can talk to your parents or guardian when you are worried. They can help you relax. They can help you let go of your worry. Letting go will help your mental and emotional and physical health.

Having a healthful attitude helps you feel good about yourself. It also makes a positive environment for others.

S. Olsson/PhotoAlto

Keep Sharp

Learning new things helps keep your mind sharp. You can learn by reading books, talking with others, and building friendships. Get plenty of physical activity, rest, and sleep, too. Your mind is alert when you are well-rested. You can think clearly. You can perform better.

Avoid Drugs

Stay away from drugs and chemicals. Some drugs can have a harmful effect on the mind. They cause your thinking to be slow and cloudy. Chemicals such as cleaning products, paint, or glue can be harmful. These products should be used in a well-ventilated area. Opening a window can help.

ACTIVITY — Math LINK

Build Your Math Memory

Write multiplication facts and their products on separate cards. Place all the cards facedown. Turn over two cards. If they show a fact and its product, pick up the cards. If not, put them back facedown and try again.

What is a healthful attitude?

LESSON REVIEW

Review Concepts

1. **Describe** one way you can share emotions in a healthful way.
2. **Define** an I-message.
3. **Name** two things you can do to stay in a healthful mood.
4. **Name** three things you can do to keep your mind healthy.

Critical Thinking

5. **Apply** Suppose your best friend is moving away. You are sad. What are some ways you can stay in a healthful mood?
6. LIFE SKILLS **Use Communication Skills** Suppose a friend borrowed a special pen from you and then lost it. Write an I-message that tells your friend how you feel.

LESSON 4

Good Character Matters

You will learn . . .

- to identify the six traits of good character.
- why your friends and heroes should have good character.
- how to show respect and what to do if you do something wrong.

Vocabulary

- **good character**, *A23*
- **hero**, *A24*
- **values**, *A24*
- **apologize**, *A27*

Have you ever wished that you could be a superhero? You can! By showing good character, you can be your own superhero every day.

Ed-Imaging/McGraw-Hill Education

Good Character

Do you tell the truth? Are you kind to others? Your actions can show that you have good character. **Good character** is acting in ways that show healthful traits, such as caring and respect. Other healthful traits are trustworthiness, responsibility, citizenship, and fairness.

If you have good character, you are trustworthy. You always tell the truth. You respect other people by treating them the way you want to be treated. You are responsible. You can be counted on to do what you say you will do. You do your part to make the world a better place. You are fair and share things evenly. You are caring and kind to others.

Good character is something you can show every day. Sometimes it is easy to show good character. Sometimes it is not easy. Showing good character takes self-control.

List the six traits of good character.

ACTIVITY

Art LINK

Make a Hangtag

Make a good character hangtag. A hangtag is a sign that can be hung on a doorknob. Cut a hangtag out of cardboard. Use a marker to write "Hang on to Good Character" at the top. Then write:

- I will tell the truth.
- I will show respect.
- I will be fair.

Hang your tag where you can see it to remind you of good character.

Do your actions at school show that you have good character?

Friends with Good Character

Write About It!

What I Admire About You Write a letter to your friend. In your letter tell your friend what parts of his or her character you admire. Tell your friend how he or she has been your hero.

Think about one of your friends. Try to remember when you first met. Did you try to learn about this person? Did you listen to what the person said? Did you watch the person's actions?

Your friend's words and actions showed his or her values. **Values** are ideas that guide the way a person acts. Some people think being kind is important. Others believe it is important to work hard.

These are examples of values. You probably make friends with people who have the same values as you. You agree on what is important. You agree on how to live your life.

A **hero** is a person you look up to because of what he or she stands for or has done. A hero should show healthful values and good character.

Who are your heroes? Do your heroes' actions show his or her values?

How do your friends show good character?

Choosing Friends

It is important to choose friends with good character. Friends with good character will suggest healthful activities for you to do together. They will not tempt you into wrong actions. How do you know if a friend has good character? Choose friends who

- practice healthful behaviors and avoid risk behaviors.
- make responsible decisions.
- say "no" when asked to do something wrong.
- tell when they have done something wrong and try to make up for it.
- do not resolve conflict by fighting.

It is not always easy to hang on to good character. Friends, though, can help each other. You and your friends can be health advocates. You can follow the steps in the activity at right.

What are two traits to look for in a friend?

ACTIVITY
LIFE SKILLS
CRITICAL THINKING

Be a Health Advocate

You can be an advocate for good character. Work with a partner and follow the steps below.

1. **Choose a healthful action to communicate.** Which trait of good character do you think is most important in school? Choose one of the traits to practice and support.
2. **Collect information about the action.** Think about ways you and your classmates can practice the trait you chose. Write them down.
3. **Decide how to communicate this information.** Write a script or a public service announcement.
4. **Communicate your message to others.** Follow through on your plan. Perform your ideas for your classmates.

Showing Respect

ACTIVITY

Social Studies LINK

Explore a World of Ways to Show Respect

People of other cultures may have different ways of showing respect. One way Japanese people show respect is to bow. Choose a culture you want to learn about. Find out how the people show respect. Act out your findings for the class.

If you have good character, your actions show respect. You can show respect for your family members by following their rules. You can show respect for your friends by helping them. You can show respect for your classmates and teachers by using manners. Others remember these actions. They know that you think highly of them. Their respect for you will grow in return.

What If I Do Something Wrong?

Everyone makes mistakes. Sometimes a mistake happens because you made an unwise choice. Sometimes a mistake happens by accident. Either way, you need to correct the mistake. You need to take action.

You can show respect by talking politely to another person.

Suppose your friend just got a new game. You borrow it and lose one of the pieces. You need to correct the mistake. First you **apologize**, or say you are sorry. Tell your friend you are sorry, and mean it. Then you take action to make it up to him. Maybe you buy a new game for your friend. Maybe you give your friend one of your games. Maybe you ask an adult to help you make a piece like the one you lost.

You have corrected your mistake. It is important to try to avoid making the same mistake again. You could try not to borrow others' things. When you have to borrow something, you take care of it.

▲ Suppose someone broke a friend's wagon. What should he or she do?

What should you do if you do something wrong?

LESSON REVIEW

Review Concepts

1. **Identify** the six traits of good character.
2. **Tell** why you might want to choose friends who share your values.
3. **Explain** how apologizing to someone when you make a mistake shows good character.

Critical Thinking

4. **Evaluate** Choose one of the six traits of good character. Why is this an important trait for a friend to have? Explain your answer.
5. LIFE SKILLS **Be a Health Advocate** You want to practice citizenship at school. Explain how you and your classmates could advocate for making school a safer place.

LESSON 5

Making Responsible Decisions

You make decisions every day. Even simple decisions, like how to spend your free time, affect your health. In this lesson you can learn how to make decisions that help you stay healthy.

You will learn . . .

- what questions to ask before you make a decision.
- ways to communicate with responsible adults about health decisions.
- how to use resistance skills.

Vocabulary

- **responsible decision**, *A29*
- **resistance skills**, *A30*

Eclipse Studios/McGraw-Hill Education

Responsible Decisions

A **responsible decision** is a choice you make that is safe and healthful and follows family guidelines. Making responsible decisions can help you stay healthy. You also make your parents or guardian proud.

Sometimes it is hard to make a responsible decision. Answering the six questions listed will help. A "no" answer to one or more of the questions shows that a decision is not responsible.

In making a hard decision, ask your parents or guardian or another responsible adult for help. Tell the adult all the facts. Tell him or her the choices that you might make. Ask the adult what he or she would do. Listen carefully, and make sure you understand the answer. Then make a responsible decision.

Tell why wearing safety gear when skating is a responsible decision.

Guidelines for Making Responsible Decisions™

1. Is it healthful?
2. Is it safe?
3. Does it follow rules and laws?
4. Does it show respect for myself and others?
5. Does it follow my family's guidelines?
6. Does it show good character?

Resistance Skills

Suppose your friend asks you to go swimming. There will not be any adults watching. You review the *Guidelines for Making Responsible Decisions*™. You answer "no" to at least one question. You decide not to go swimming with your friend. How can you tell him your decision? You can use resistance skills. **Resistance skills** are ways to say "no" to unwise decisions.

Resisting Risk Behaviors

When you use resistance skills, you say "no" in a firm voice. You also give reasons for your decision. One reason might be that swimming with no adults watching breaks a rule. Healthful behaviors follow rules. Risk behaviors go against rules.

Write About It!

How to Say "No" Write a "How To" paragraph listing the steps used to say "no." Include what you should do at each step. Share your writing with the class.

Make sure your actions match your words when you say "no."

Next you act on your words. You know it is unsafe to swim alone, so you decide to ride your bike instead. You do not let your friend talk you into swimming. It also is important to tell an adult about his unwise decision.

What are resistance skills?

ACTIVITY

LIFE SKILLS

CRITICAL THINKING

Use Resistance Skills

Work with a partner to practice using resistance skills. Suppose a classmate wants to copy your answers during a test. First, write about how you would say "no." Include reasons you would give for your decision. Then, role-play the situation with your partner. Follow the steps for using resistance skills.

1. **Look at the person. Say "no" in a firm voice.**
2. **Give reasons for saying "no."**
3. **Match your behavior to your words.**
4. **Ask an adult for help if you need it.**

LESSON REVIEW

Review Concepts

1. **List** six questions to ask when you have to make a hard decision.
2. **Explain** how to ask for help when making a responsible decision.
3. **Describe** when you should use resistance skills.
4. **List** the steps to use to say "no."

Critical Thinking

5. **Predict** Your friend wants you to walk to his house but it's dark. Use the *Guidelines for Making Responsible Decisions*™. What might happen if you choose to go? What might happen if you use resistance skills instead?
6. LIFE SKILLS **Use Resistance Skills** Suppose that when telling a friend "no," you begin to laugh. Do you think your friend will believe your words? Tell why or why not.

Learning LIFE SKILLS

CRITICAL THINKING

Make Responsible Decisions

Problem María's friend wants her to take a shortcut through an unsafe place. What should she do?

Guidelines for Making Responsible Decisions™

- Is it healthful?
- Is it safe?
- Does it follow rules and laws?
- Does it show respect for me and others?
- Does it follow family guidelines?
- Does it show good character?

Solution María has the responsibility of staying safe. She decides to use the four steps on the next page to make a responsible decision.

Foldables™ To Learn Life Skills

Learn This Life Skill

Follow these steps to make responsible decisions. The Foldables™ can help.

1 Identify your choices. Check them out with your parent or trusted adult.

Maria can go with her friend or say "no."

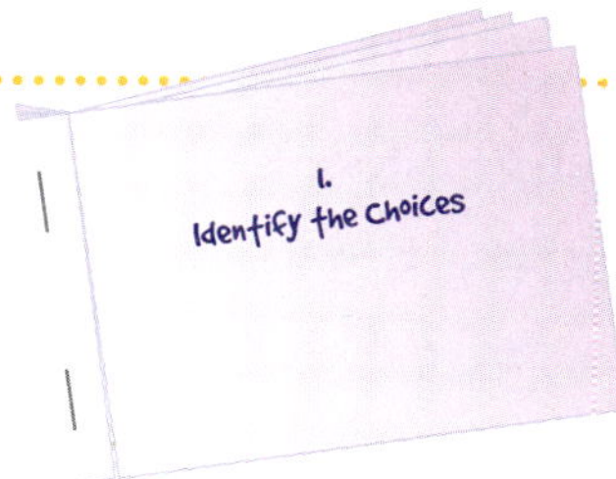

2 Evaluate each choice. Use the *Guidelines for Making Responsible Decisions™*.

Maria answers the six questions on page A32 for making responsible decisions.

3 Identify the responsible decision. Check this out with your parent or a trusted adult.

Maria said "no" to at least one of the questions. That means taking the shortcut is not wise.

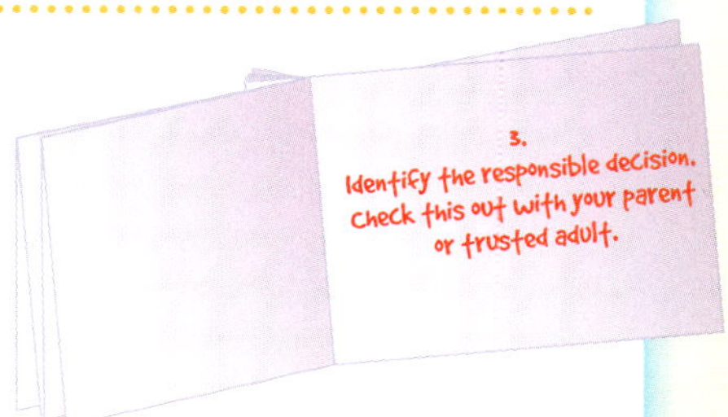

4 Evaluate your decision.

Maria explains her decision to her friend.

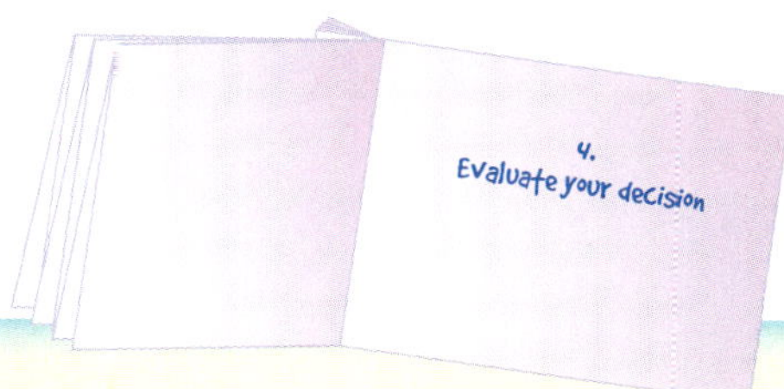

Practice This Life Skill

Write down a time when you made a hard decision. Swap with a classmate. Make a responsible decision using the steps above.

LESSON 6

Manage Stress

You will learn . . .

- what effect stressors have on your body.
- the difference between healthful and harmful stress.
- ways to manage your stress.

Vocabulary

- **stress**, *A35*
- **stressor**, *A35*
- **healthful stress**, *A36*
- **harmful stress**, *A37*

The people in your life expect a lot from you. You also expect a lot from yourself. You want to be the best person you can be. Sometimes it is hard to do everything well. Sometimes trying too hard affects your health.

Stressors

Stress is the way your body reacts when there are changes in your life. Everyone feels stress. It is a natural part of life. That's because you deal with stressors every day. A **stressor** is something that causes stress. Every person has different stressors. Talking in front of the class might be a stressor for you. It might not bother someone else. The sound of a baby crying might not cause you stress. This sound, though, probably is a stressor to the baby's parents or guardian.

Stress causes changes in your body. These changes can help you get ready for quick action. Look at the photo of the student. He feels stress. Read about what is happening to his body because of stress.

What is a stressor?

Do You Know?

Good changes can cause stress. They cause your body to react. Getting an "A" on a test, adopting a new kitten, or buying a new skateboard might be stressors.

Effects of Stress

- **Your heart may beat faster to pump blood to your muscles.**
- **You may breathe faster to get more oxygen.**
- **Sugar stored in your body enters the blood to give you more energy.**
- **Your hands get sweaty.**
- **Your knees feel shaky.**

The Two Types of Stress

Stress can be healthful. **Healthful stress** is stress that helps you perform well. It also helps keep you healthy.

Suppose you are a pitcher on a baseball team. Your team will win the game if you strike out the next batter. You feel stress. You breathe faster to get more oxygen. Your heart beats faster. Your body releases sugar into your blood. This sugar gives you energy. You throw the ball faster than ever before. The batter strikes out, and your team wins the game.

Your body changed due to stress. These changes gave you energy and helped you perform better. The stress was healthful.

What may be a stressor to one person might not be a stressor to another person. Why might that be?

Harmful Stress

Sometimes stress is harmful. **Harmful stress** is stress that harms health or causes you to perform poorly.

Suppose you are in a school play. You have not practiced your lines. You are worried about how you will do. You have trouble falling asleep. On the day of the performance you get a stomachache. Just before the play starts, your head starts to hurt. You cannot remember your lines.

Harmful stress caused changes in your body. You had trouble sleeping. You got a stomachache and a headache. If the stress continues, it could lead to illness.

What are some ways that harmful stress affects the mind and body?

CAREERS

School Counselor

School counselors help students when they are in school. They can help students with many kinds of problems. For example, they help students manage stress. They also help students plan for their future.

School counselors must have at least a four-year college degree. They may have even more training. They must be certified.

If you are interested in helping people, think about being a school counselor.

Students who feel stress can get help from counselors at school. Many youth clubs and community centers have counselors who can help.

Managing Stress

Manage Stress

1. **Identify the signs of stress.**
2. **Identify the cause of stress.**
3. **Do something about the cause of stress.**
4. **Take action to reduce the harmful effects of stress.**

Whenever you experience stress, be aware of the signs. Then identify what you think is causing the stress. Try to do something about the cause. Failing to practice lines for the school play can cause stress. You can talk to your parents or guardian about stress. They can help you decide what to do about the cause of stress. You may not know what is causing stress. They can help you figure it out.

▲ **Getting plenty of sleep is one way to manage stress.**

Do You Know

When you feel stress, talk to parents or guardians. Other trusted adults to talk to are

- teachers
- school counselors
- school nurse
- family doctor
- health clinic health professionals
- community center professionals
- youth club counselors

Reducing Stress

You can take action to reduce the harmful effects of stress.

- Get plenty of rest and sleep.
- Get plenty of physical activity each day.
- Eat healthful meals and snacks.
- Take slow, deep breaths. This slows down your heart rate and blood pressure.
- Take time during the day to relax.

Describe two actions to manage harmful stress.

ACTIVITY LIFE SKILLS **CRITICAL THINKING**

Manage Stress

Your book report for school is due tomorrow. You haven't finished reading the book yet. You have a stomachache and feel tired. Work with a partner. Act out ways to reduce stress for your class.

1. **Identify the signs of stress.** What signs show you are under stress? Act these signs out.
2. **Identify the cause of stress.** Talk to your partner about what's causing this stress.
3. **Do something about the cause of stress.** With your partner, make a plan that would help you get your report done.
4. **Take action to reduce the harmful effects of stress.** With your partner, determine how you can reduce stress and still get your report done. Act out your ideas for the class.

LESSON REVIEW

Review Concepts

1. **List** ways your body changes because of stress.
2. **Contrast** healthful stress and harmful stress.
3. **Explain** how getting rest, getting physical activity, and eating healthful foods can help you manage the effects of stress.

Critical Thinking

4. **Analyze** How might a school counselor help you manage stress?
5. LIFE SKILLS **Manage Stress** A friend is feeling the effects of harmful stress. What are some things she can do to reduce the effects of stress?

CHAPTER 1 REVIEW

Use Vocabulary

attitude, *A20*
good character, *A23*
harmful stress, *A37*
health goal, *A8*
respect, *A12*
responsible decision, *A29*

Choose the correct term from the list to complete each sentence.

1. Treating others as you want to be treated is __?__.
2. The way you think, act, or feel is your __?__.
3. Something you work toward to help you become a healthier person is a(n) __?__.
4. Stress that can cause you to perform poorly is __?__.
5. A safe and healthful choice is a(n) __?__.
6. Acting in ways that show healthful traits, such as caring and respect, is __?__.

Review Concepts

Answer each question in complete sentences.

7. What is an example of a risk behavior?
8. What does the first part of an I-message describe?
9. What are three important things to do when you make a mistake?
10. How can healthful stress help you perform well?

Reading Comprehension

Answer each question in complete sentences.

Doing your best includes doing your best in school. People study and learn in different ways. Some people learn best by listening. Some people learn best by reading. Others learn best by writing things down. These different ways to learn are called learning styles.

11. What is a learning style?
12. What are three ways people learn?
13. How can discovering your learning style help you do your best?

Critical Thinking/Problem Solving

Answer each question in complete sentences.

Analyze Concepts

14. Why is it important to care for the three parts of health?

15. Explain how to make a health behavior contract.

16. How does using communication skills show good character?

17. How can you have a good attitude about studying?

18. How can you show family members that you are fair?

Practice Life Skills

19. **Use Communication Skills** A friend promised to help you study for a test. The day before the test, he called and said he was going to a movie instead. You are angry. Write an I-message to your friend about how you feel.

20. **Make Responsible Decisions** Think about a hard decision you have made or need to make. List the *Guidelines for Making Responsible Decisions*™ and your answer to each question. Describe the decision you made. Give reasons for your choice.

Read Graphics

Use the chart to answer questions 21–23.

Derek's Ways to Show Good Character

Quality	Action
Trustworthiness	Always tell my parents the truth about where I am going.
Respect	Ask before I borrow from someone.
Responsibility	Turn in my homework on time.
Citizenship	Do something to help in my community.
Fairness	Let my friends play with my toys and games.
Caring	Do something to help children or older adults.

21. Name one other action Derek could include to show trustworthiness.

22. Derek wants to start a food drive project in his class. Which quality would this action be listed as?

23. Derek reads a book to his younger sister at night. Which qualities might this action relate to?

CHAPTER 2

Family and Social Health

What Do You Know?

People who are in healthful relationships respect each other. Read each statement. Write **yes** if you agree. Write **no** if you disagree.

Respect means:

___?___ treating others as you would like to be treated.

___?___ helping your family with chores.

___?___ listening to your friends ideas.

Did you answer **yes** to each statement? Great! All these statements tell ways to show respect. To learn more about healthful relationships read **Family and Social Health**.

LESSON 1

Healthful Relationships

You will learn . . .

- ways to make a healthful relationship.
- ways to show respect for other people.
- how to communicate in healthful ways.

Vocabulary

- **relationship**, *A45*
- **respect**, *A45*
- **communicate**, *A47*
- **body language**, *A47*

Think about the people in your life. You have family members and friends. You have teachers and classmates. You have a relationship with all those people.

Izabela Habur/Getty Images

Relationships and Respect

A **relationship** is a connection you have with another person. You have many kinds of relationships. You have relationships with family members, friends, teachers, classmates, and neighbors.

In a healthful relationship, there is respect. **Respect** is treating others as you want to be treated. There also is communication.

Respecting Other People

You can show respect in these ways.

- **Show good character.** Do not ask others to do anything wrong. Do not ask friends to break school rules. Encourage your friends to show good character.
- **Follow family guidelines.** Show respect for your *family guidelines,* or rules set by parents or guardians that tell you how to act. Do not break rules. Don't do something you know you shouldn't do.
- **Help others stay healthy.** Encourage friends to eat healthful snacks instead of junk food, for example.

BUILD Character — ACTIVITY

Make a Chart

Respect Make a four-column chart. Write "Family," "Friends," "Teachers," and "Self" across the top of each column. List five actions that show respect for your family, friends, teachers, and yourself. Hang the chart where you can see it. Try to show respect everyday in these ways.

Write About It!

Communicate with Respect Write a paragraph about showing respect to your teacher when you communicate. Present your paragraph to the class. Show how you are respectful when you communicate to your teacher.

- **Choose actions that keep people safe.** Follow safety rules in the classroom and on the playground.
- **Avoid violence.** *Violence* is harm done to yourself, others, or property. Kicking, pushing, or punching is violence. Taking things from others is violence.

▼ Helping family members shows respect. In what other ways might you show respect at home?

Getting Respect from Other People

In healthful relationships, other people respect you. Here are some ways to get respect.

- **Have self-respect.** *Self-respect* is thinking highly of yourself. When you show self respect, others will show respect for you.
- **Set limits.** To *set limits* is to say what actions are OK and not OK with you.
- **Use honest talk.** Tell others how you feel. Maybe someone did not show respect for you. You might feel angry or hurt.
- **Ask others to change disrespectful action.** If someone shoves you, tell him or her not to do it ever again.

When people respect each other they can work together better. When you do the activities in this book with your classmates, follow all the ways of giving and getting respect. You'll find you can do the activities faster and better.

What is a relationship?

Healthful Communication

Healthful relationships are built on good communication. When you **communicate**, you share feelings, thoughts, or information. There are two parts to communication: sending messages and listening.

Three Ways to Send Messages

Good communication is getting your message across.

- **Speaking and Writing** You might speak to someone or send an email. You could write a note, or draw a picture. When talking with someone, be polite. Use correct grammar. Give the person your full attention. Use I-messages to say how you feel.
- **Body Language** The way you move your body to communicate is called **body language**. The way you use your hands and arms sends a message to others. The look on your face and the way you sit and stand sends a message, too.
- **Sign Language** People who cannot hear use their hands to communicate. This is called *sign language.*

Write About It!

Ways to Say "I'm Sorry" Think of a time when someone you know hurt a friend. Maybe the person teased his or her friend. The friend's feelings were hurt. Work with a partner. Write several ways to say "I'm sorry" such as "I apologize for being mean." Then take turns role-playing your ideas.

BUILD Character — ACTIVITY

Respect Our Teacher

Respect Do a role-play with a group. Show how to use respect when group members communicate with your teacher.

MAKE a Difference

Make Grand-friends

Christina of Corunna, Michigan adopts "grand-friends." She visits elders at the Shiawassee County Medical Care Facility. They talk, play games, and read books together. Christina's visits cheer the elders. She thinks it's important to do something nice for others.

Two Ways to Show You Are Listening

When you listen, you can show that you hear and understand.

- **Give the other person your full attention.** When the other person speaks, look directly at him or her. Let the person finish before you begin to talk. If you are talking on the telephone, don't watch television at the same time.
- **Pay attention to body language.** Remember, your body language sends messages. Are you smiling when you listen? Are you frowning? Are you tapping your foot? What messages do these actions send? Be aware of the other person's body language, too. Is the person interested or bored?

Tell some healthful ways to share what you feel, think, and know.

▶ **When listening to someone, give the person your full attention. Look at her.**

ACTIVITY

LIFE SKILLS

CRITICAL THINKING

Use Communication Skills

Using your own words to repeat what another person says helps communication. This is called *paraphrasing*. Work with a partner. Role-play this skill.

1. **Choose the best way to communicate.** Say or read a sentence to your partner. You might want to choose a sentence from a book.
2. **Send a clear message. Be polite.** Speak clearly. Look at your partner.
3. **Listen to each other.** Have your partner paraphrase what you said. How did your partner do?
4. **Make sure you understand each other.** Talk about how well you communicated. Did your partner understand you? How could you make what you said clearer? How could your partner listen better? Switch roles and try the activity again.

LESSON REVIEW

Review Concepts

1. **List** three ways you might show respect for others.
2. **Describe** three ways you might get respect from others.
3. **Define** the term body language.

Critical Thinking

4. **Analyze** You do not ask your friends to do something wrong. How does this show respect?
5. LIFE SKILLS **Use Communication Skills** You tell Nathan to meet you near the swings. Nathan looks for you near the slide. What might have caused the mix up?

LESSON 2

Resolving Conflict

You will learn . . .

- how to apply the four steps to resolve conflict.
- ways to get along with others.
- how peer pressure works.

Vocabulary

- **conflict**, *A51*
- **peer**, *A54*
- **peer pressure**, *A54*

Sometimes people disagree. This can lead to conflict. You can learn to resolve conflicts without fighting.

How to Resolve Conflict

A disagreement is a **conflict**. Some conflicts are easy to handle. Other conflicts might be harder to handle. They might include angry feelings. There are four steps you can use to resolve conflict. Using the steps shows respect and keeps relationships healthful.

Resolve Conflicts

1. Stay calm.
2. Talk about the conflict.
3. List possible ways to settle the conflict.
4. Agree on a way to settle the conflict. You may need to ask a responsible adult for help.

Steps to Resolve Conflict

1. **Stay calm.** Do not act on any strong feelings.
2. **Talk about the conflict.** Use I-messages to tell what you are feeling. Allow time for each person to speak.
3. **List possible ways to settle the conflict.** Work together to find ways to resolve the conflict. Check out each way to settle the conflict using the *Guidelines for Making Responsible Decisions*™.
4. **Agree on a way to settle the conflict. You may need to ask a responsible adult for help.** Agree on the best solution. If you need help, ask a responsible adult.

Why are some conflicts harder to resolve than others?

Getting Along with Others

Being fair means following rules, taking turns, and sharing. Being fair helps you get along better with others. When you are not fair, conflict can happen.

Ways to Get Along with Others

Here are some other actions you can take.

- **Think about your actions.** How will they affect other people? Put yourself in their shoes. How would you feel?
- **Show respect.** Do not use or borrow items without asking.
- **Share.** Take only your portion. Share what you have with others.
- **Take turns.** Taking turns is one of the easiest ways to be fair.
- **Play by the rules.** Follow family guidelines and school rules. Do not cheat when playing games or doing homework.
- **Do not *gossip*, or say unkind words, about others.** Gossip can hurt other people's feelings. It can also change the way people see you. They might think you can't be trusted.

▼ When playing on a team, always play fair. Follow the rules.

The Harmful Effects of Gossip

When you gossip you change the way people think about you. Some people may think that if you gossip about others, you gossip about them, too. Here is what can happen if you gossip.

- **People will not trust you if you gossip.** Suppose a friend tells you something that upset her. You tell other classmates. If your friend finds out, she will no longer trust you. You could even lose her as a friend.
- **People will have hurt feelings if you gossip.** Suppose a friend gets a new haircut. You make fun of his new haircut to another friend. He hears what you said. He now has hurt feelings.
- **People can get angry.** Suppose you tell others something that isn't true about a classmate. How do you think she might feel? She might feel hurt. She might get angry. She might try to get even with you.
- **People might think you are disrespectful or a coward.** Suppose a classmate does something that you do not like. You tell others, but you don't tell him. He finds out what you said. He might think you are a coward. You did not tell him how you felt.

What is gossip?

ACTIVITY — Con$umer Wi$e

Go to the store with your parents or guardian. Look for games and activities in which players work together. You might see competitive games. *Competitive* means someone wins, and someone loses. You will also see activities, such as games and puzzles, where people work together. When people work together, they *cooperate*. Write down the cooperative games you find.

Activity

Health Online

Peer Pressure

Research peer pressure to find out how it influences the decisions people make. Use the e-Journal writing tool to write a report on your findings.

What to Do About Peer Pressure

A **peer** is someone who is the same age as you. Sometimes peers try to get you to act or think a certain way. The effect that other children your age have on you is called **peer pressure**.

There are two types of peer pressure: positive and negative. Positive peer pressure is healthful. For example, friends might influence you to do your homework on time.

Negative peer pressure can be harmful. Being pressured to cheat or lie are examples of negative peer pressure. Avoid peers who try to use negative peer pressure against you.

When Someone Wants to Fight

A *fight* is a struggle between two or more people. A fight might be with words. It might include yelling. A fight might involve punching, shoving, and kicking. A peer who threatens or frightens others is a *bully*. People who are bullied often feel alone, very sad, and confused.

If someone wants to fight, say that you do not want to fight. Get away from the situation. If the person tries to harm you, ask a responsible adult for help.

What should you do if someone tries to get you to fight?

LIFE SKILLS
ACTIVITY
CRITICAL THINKING

Analyze What Influences Your Health

True friends influence you in healthful ways. They help you build good character. They help you succeed. How do your true friends influence you?

1. **Identify people and things that can influence your health.** Name a true friend who influences your health.
2. **Evaluate how these people and things can affect your health.** Write a paragraph. Tell how this true friend influences you. How does this friend use peer pressure?
3. **Choose healthful influences.** Write another paragraph. Tell why this friend is a good choice.
4. **Protect yourself against harmful influences.** Knowing how much influence others have on your health can keep you healthy. Work with a classmate. Role-play ways to stay away from peers who use negative peer pressure to influence your choices.

LESSON REVIEW

Review Concepts

1. **List** the four steps you can use to resolve conflict.
2. **Explain** ways to get along with others.
3. **Identify** the two kinds of peer pressure.

Critical Thinking

4. **Conclude** How might a responsible adult help you deal with a bully?
5. LIFE SKILLS **Analyze What Influences Your Health** You invite a friend to join your soccer team. How is this a healthful influence?

Learning LIFE SKILLS

CRITICAL THINKING

Resolve Conflicts

Problem After the teacher showed the class a book, Thomas and Charlie went to the bookshelf for the same book. They find only one copy on the shelf. Each boy wants it.

Solution What could Thomas and Charlie do to settle their conflict? Use the four steps on the next page.

Foldables™ To Learn Life Skills

Learn This Life Skill

Follow these four steps to resolve conflict. The Foldables™ can help you.

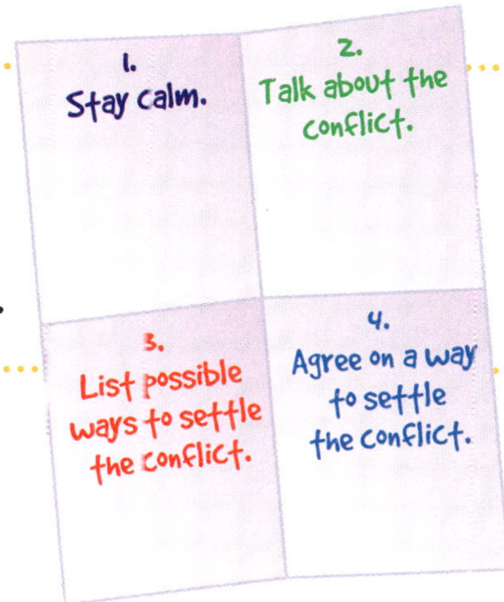

1 Stay calm.

Before becoming too angry, the boys should calm down. They begin to look at the situation.

2 Talk about the conflict.

The boys use I-messages to talk. They come up with ideas for resolving the conflict.

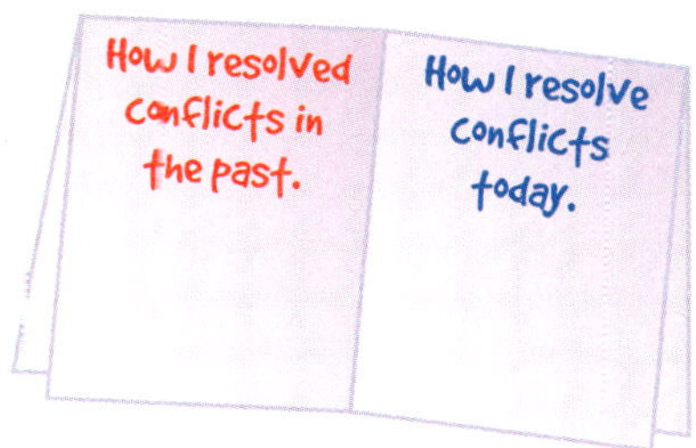

3 List possible ways to settle the conflict.

The boys check out each way to settle the conflict. They use the *Guidelines for Making Responsible Decisions*™.

4 Agree on a way to settle the conflict. You may need to ask a responsible adult for help.

The boys go to their teacher. They ask whether other copies of the book are available.

Practice This Life Skill

Work in groups of four. Think of a conflict that could happen at school. Write a skit to resolve the conflict. Use the four steps above. Perform the skit for the class.

LESSON 3

My Family

You will learn . . .

- what it means to be a family.
- ways to be a responsible family member.
- how to adjust to family changes in healthful ways.

Vocabulary

- **family**, *A59*
- **memories**, *A59*
- **value**, *A60*

What do you think of when you hear the word *family*? Some families keep a photo album. It includes pictures of family members as they grow and change. Then family members can look at the photos and remember past times.

What Is a Family?

A **family** is a group of people who are related in some way. Families can be different sizes. They can be made up of different family members.

- A *nuclear family* is made up of a husband and a wife and one or more children.
- A *single-parent family* has only one parent with one or more children.
- A *blended family* may include a husband and wife who were married before. They each may have children from a previous marriage. They may have children together.
- An *extended family* may include grandparents, aunts, uncles, and cousins. They may even live with you.

Family members care about one another. They love, support, and trust each other. Their **memories**, or things remembered from the past, keep them close for many years.

Name three types of families.

ACTIVITY

On Your Own

FOR SCHOOL OR HOME

Write About a Favorite Photo

Find a family photo that has meaning for you. Write two or three sentences telling what memories you have about the photo.

Families spend time together.

Family Values

Values are learned in families. **Values** are ideas that guide the way a person acts. Your family influences your values. Suppose nutrition and fitness are important values to your parents or guardian. They might teach which foods are healthful. They might also teach you to be physically active.

Your parents or guardian teach you values. They want you to become loving and responsible. They want you to learn to care for yourself and others. They want you to be the best person you can be.

► **You learn values when you share time working with your family.**

Ways to Be a Responsible Family Member

Being a responsible family member takes practice. Here's what you can do.

- **Show respect.** Act in ways that show respect. Give family members privacy when they need it. Do not say mean things or talk back.
- **Be responsible.** Other family members can depend on you when you are responsible. Try to do your chores on time and follow family guidelines.
- **Be understanding.** Care about how your actions affect other family members. Turn down your music when a family member has a headache. Help a younger brother or sister with his or her homework.
- **Use self-control.** *Self-control* is being able to stop doing something you should not do. Follow family guidelines and stick to family values.

What are family values?

ACTIVITY LIFE SKILLS

CRITICAL THINKING

Access Health Facts, Products, and Services

Your mother has a migraine headache. You want to show her you care. Use this activity to access health facts, products, and services.

1. **Identify when you might need health facts, products, and services.** Why might you need health facts?
2. **Identify where you might find health facts, products, and services.** Talk with a parent or guardian about migraine headaches. Where might you look to find information about migraine headaches?
3. **Find the health facts, products, and services you need.** Find all the health facts you can about migraine headaches. What can you do to help ease your mother's pain? Write down the facts.
4. **Evaluate the health facts, products, and services.** Review with your parents or guardian the health facts you've found. Is the information helpful? How might you use it in the future?

Changes in Families

Every family changes. A baby may be born. A child may be *adopted* or taken into the family from another family. Parents may separate, divorce, and remarry. A family may move. A family member may be away in the military. A loved one may become ill. A family member may die.

▲ A new baby takes much of a family's time. How could you help out?

Adjusting to Change

Family changes are not always easy. But families can find healthful ways to adjust.

- **Ask others for support.** Let family members know what you think. Share how you feel. Ask your family for help.
- **Offer help.** Help others with their chores. Take time to talk.
- **Ask questions.** Adults in your family can answer your questions. They can explain what to expect when things change.
- **Manage stress.** Follow the steps to manage stress. Get plenty of rest and sleep. Eat healthful foods.
- **Show kindness to other family members.** Acts of kindness make everyone feel better.

Adjusting to Death

One of the hardest family changes to adjust to is death. A family member may die. A family pet may die. There is a loss. You may feel great sadness. You may feel angry. You may feel grief. *Grief* is feeling very sad. If you feel grief, talk to a parent or guardian. Tell how you feel. Express your emotion healthfully.

There are healthful ways to adjust to death. Share your feelings with your parents, guardian, or other family members. Remember that it is okay to feel sad or angry. Express your feelings in healthful ways. Use I-messages if you are angry. Share your memories with others.

CAREERS

Day Care Teacher

Day care teachers work with preschool children. They read stories to the children and teach them games. They show them how to get along with others. They talk about how to handle emotions. Day care teachers may also need to help children deal with changes.

What is one of the hardest changes to deal with?

LESSON REVIEW

Review Concepts

1. **Explain** what a family is.
2. **Name** three ways to be a responsible family member.
3. **Tell** three ways you can adjust to family changes.

Critical Thinking

4. **Conclude** Why do you think adjusting to death is so difficult?
5. LIFE SKILLS **Access Health Services** If a loved one dies, who might you go to for help?

LESSON 4

My Friends

You will learn . . .

- why you need friends.
- ways to make new friends.
- how friends can help each other make responsible decisions.

Vocabulary

- **friend**, *A65*
- **disability**, *A67*

You may have friends of different ages. You may enjoy talking with an older friend who knows and understands you. You may like playing with a younger friend who can be silly with you. You may like to play games or study with friends your age.

Why Do You Need Friends?

▲ Friends may share secrets with each other.

A **friend** is a person who likes and supports you. Some people have many friends. Some people may have only one or two close friends. Whatever the case, you need friends for many reasons.

- Friends teach you how to get along with others.
- Friends help you when you feel stress.
- Friends enjoy physical activities with you.
- Friends listen to your ideas and encourage you to try them out.
- Friends protect you from harm. They want what's best for you.
- Friends give you the chance to show that you are a caring person. They spend time with you. They get to know who you are.

Everyone has friends. But not all friends are true friends. How can you tell a true friend from one who isn't? A true friend is responsible and caring. A true friend wants what's best for you.

Who is a friend?

Write About It!

A True Friend Pick one true friend. Write a short paragraph. Tell what makes this person a true friend. What activities do you share together? What characteristics do you admire in your friend? Why? How would you feel if this person moved away?

Making Friends

Making new friends may seem hard. But if you follow these tips, it might be easier.

- **Think about places you can meet people.** With your parents or guardian, go places where your peers might be. You might join a sports team or a club. There may be children there that you like.
- **When you meet someone you like, make plans together.** Get permission from your parents or guardian first. Ask your parents or guardian to take you to a park where you can play with your new friend.
- **Check out the person. Make sure that he or she is caring and responsible.** It takes time to make a friend. You must spend your time wisely. Do you share the same family values as he or she does? Does this person show respect for you and others?
- **Put time and effort into being a true friend.** To make a true friend you need to be a true friend. Treat the other person the way you want to be treated. Be responsible and caring. Show respect at all times.

▲ **You can make friends with a new student. Make him or her feel welcome.**

Friends with Special Needs

You may make friends with someone who has special needs. He or she may have a disability. A **disability** is something that changes a person's ability to do certain tasks.

Suppose that your new friend uses a wheelchair. Having to use a wheelchair causes your friend to have special needs. Your friend might use a ramp to enter a building. Your friend might need someone to hold open a door. Be a good friend to someone who has special needs. Talk about ways to meet the special needs your friend has. Plan activities that your friend can do with you. Your friend in a wheelchair might be able to play basketball. Know that most of your friend's needs are the same as yours.

ACTIVITY

Social Studies LINK

Make a Poster

Select a country. Find out what health issues the people who live there experience. Make a poster telling what health issues you found. Write suggestions for how the health issue might be handled. Hang your poster in the classroom.

What is a disability?

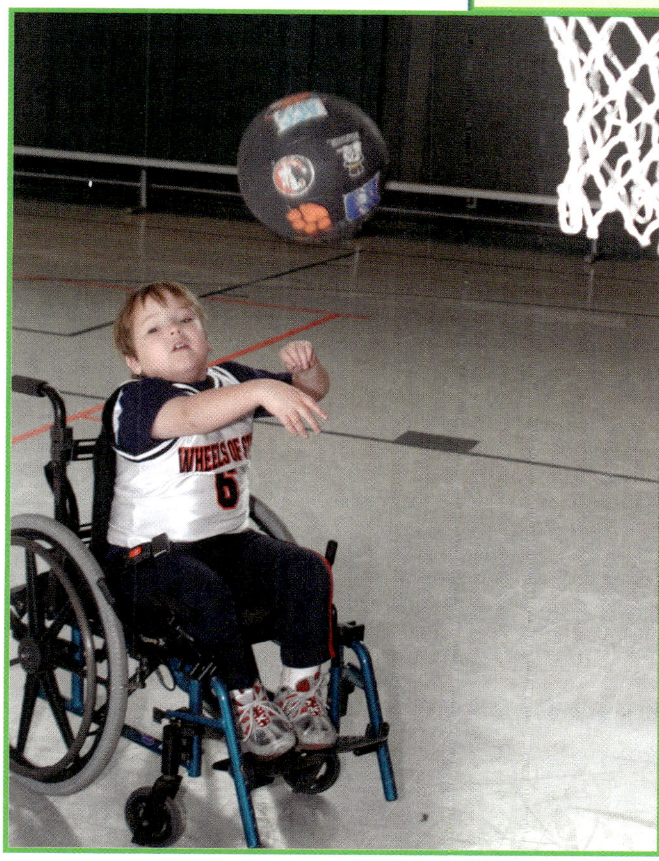

▶ Students with a disability may enjoy many of the same activities as you.

Friends Practice Making Responsible Decisions

Guidelines for Making Responsible Decisions™

- **Is it healthful?**
- **Is it safe?**
- **Does it follow rules and laws?**
- **Does it show respect for me and others?**
- **Does it follow family guidelines?**
- **Does it show good character?**

True friends help one another make responsible decisions. Here's how.

- **Use the *Guidelines for Making Responsible Decisions™*.** The six questions help you know whether a decision is responsible.
- **Do not go along with a friend's wrong decision.** Get away from the situation if it could be harmful.

Taking Responsibility for Health

True friends encourage each other to take responsibility for their health at all times.

- **Make a health behavior contract.** Suppose you and your friend want to get more physical activity. You can make a health behavior contract to help you both reach this goal.
- **Take turns choosing a healthful behavior.** You might choose ice skating one day. Your friend might choose reading. By taking turns you will always do something different.

What should you do if a friend is about to make a wrong decision?

▲ **Friends help each other be the best that they can be.**

ACTIVITY

LIFE SKILLS

CRITICAL THINKING

Make Responsible Decisions

You have made a new friend. You plan to ride bikes after school. You aren't sure where to ride. What should you do?

1. **Identify your choices. Check them out with your parent or trusted adult.** You could ride near your house. Where else might you ride bikes?
2. **Evaluate each choice. Use the *Guidelines for Making Responsible Decisions*™.** Answer the questions for each choice.
3. **Identify the responsible decision. Check this out with your parent or trusted adult.** Explain why your decision is responsible.
4. **Evaluate your decision.** Explain what might happen if you make the decision. Analyze your decision. Is it responsible? If not, go back to step 3.

LESSON REVIEW

Review Concepts

1. **Name** three reasons you need friends.
2. **Tell** three ways to make friends.
3. **Identify** three ways you can make responsible decisions with friends.

Critical Thinking

4. **Analyze** When might you need to make a new friend?
5. LIFE SKILLS **Make Responsible Decisions** You are skateboarding with a friend. She skates down a busy street. What should you do?

CHAPTER 2 REVIEW

Use Vocabulary

- **body language,** *A47*
- **conflict,** *A51*
- **disability,** *A67*
- **friend,** *A65*
- **memories,** *A59*
- **peer,** *A54*
- **relationship,** *A45*
- **value,** *A60*

Choose the correct term from the list to complete each sentence.

1. A disagreement is a(n) __?__.
2. Someone who is the same age as you is a(n) __?__.
3. The way you move your body to communicate is __?__.
4. A person who likes and supports you is a(n) __?__.
5. Things that you remember from your past are __?__.
6. A connection you have with another person is a(n) __?__.
7. Something that changes a person's ability to do certain tasks is a(n) __?__.
8. A(n) __?__ is an idea that guides the way a person acts.

Review Concepts

Answer each question in complete sentences.

9. What is peer pressure?
10. What is the first step to use to resolve conflict?
11. What is a value?
12. What is the meaning of the word *communicate*?
13. What is a disability?

Reading Comprehension

Answer each question in complete sentences.

- **Show good character. Do not ask others to do anything wrong. Do not ask friends to break school rules. Encourage your friends to show good character.**
- **Follow family guidelines. Show respect for your *family guidelines*, or rules set by parents or guardians that tell you how to act. Do not break rules. Don't do something you know you shouldn't do.**

14. Why shouldn't you ask others to do something wrong?
15. What are family guidelines?

Critical Thinking/Problem Solving

Answer each question in complete sentences.

Analyze Concepts

16. You're at home watching your favorite television program. Your mom comes home from the grocery store. What might you do to show you care? Explain how your actions show caring.

17. You are playing baseball. You notice that a teammate is cheating. What might you do?

18. At lunch two of your friends are gossiping about another friend. You don't participate. Your friends become angry with you. Did you do the right thing? Why or why not?

19. Why might a peer bully other people?

20. List the four steps in making responsible decisions. How might these steps help you decide if a friend is a true friend?

Practice Life Skills

21. **Resolve Conflicts** A friend gets angry with you because you did not play with her during recess. How could you resolve this conflict?

22. **Make Responsible Decisions** Your younger brother wants to play football with you and your friends. You are worried that he may get injured. He is so much smaller than everyone else. Use the *Guidelines for Making Responsible Decisions*™ to help you make a responsible decision.

Read Graphics

Complete the chart below by listing the type of family for each description.

Types of Families	
23.	One parent with two children
24.	Two parents, one child, a grandparent, and an uncle
25.	Two parents and three children
26.	Two parents who were married before with children of their own. One new baby

UNIT A ACTIVITIES AND PROJECTS

Effective Communication

Write a Letter

Your family has moved to a new city. You left your best friend behind. Write a letter to your friend telling how you feel. Include in your letter ways the two of you can stay in touch.

Self-Directed Learning

Make a Family Time Line

Bring in family photos to show how your family has changed. For each photo, write a caption telling about the picture.

Critical Thinking and Problem Solving

Write a Skit

Work with a partner and write a skit about family guidelines. For each rule you include, determine what might happen if it is broken. Perform the skit for your class.

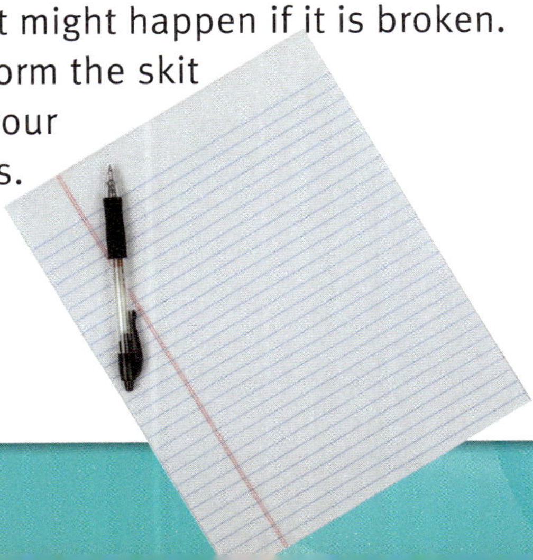

Responsible Citizenship

Adopt a Senior Citizen

Plan a class project to adopt a group of senior citizens. Make cookies or write notes and send them to the group. Think of other activities you might be able to do with your new friends.

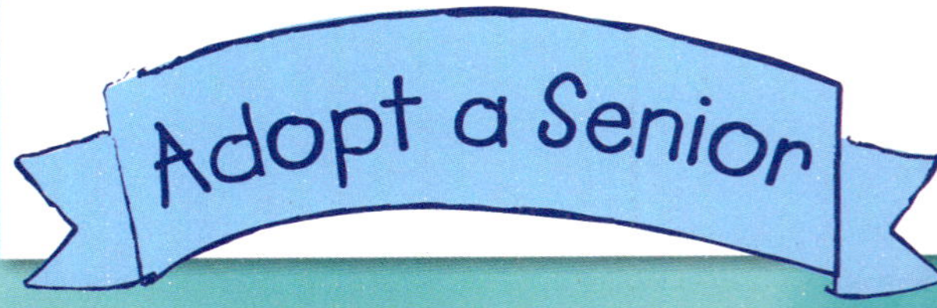

UNIT B

Growth and Nutrition

CHAPTER 3

Growth and Development

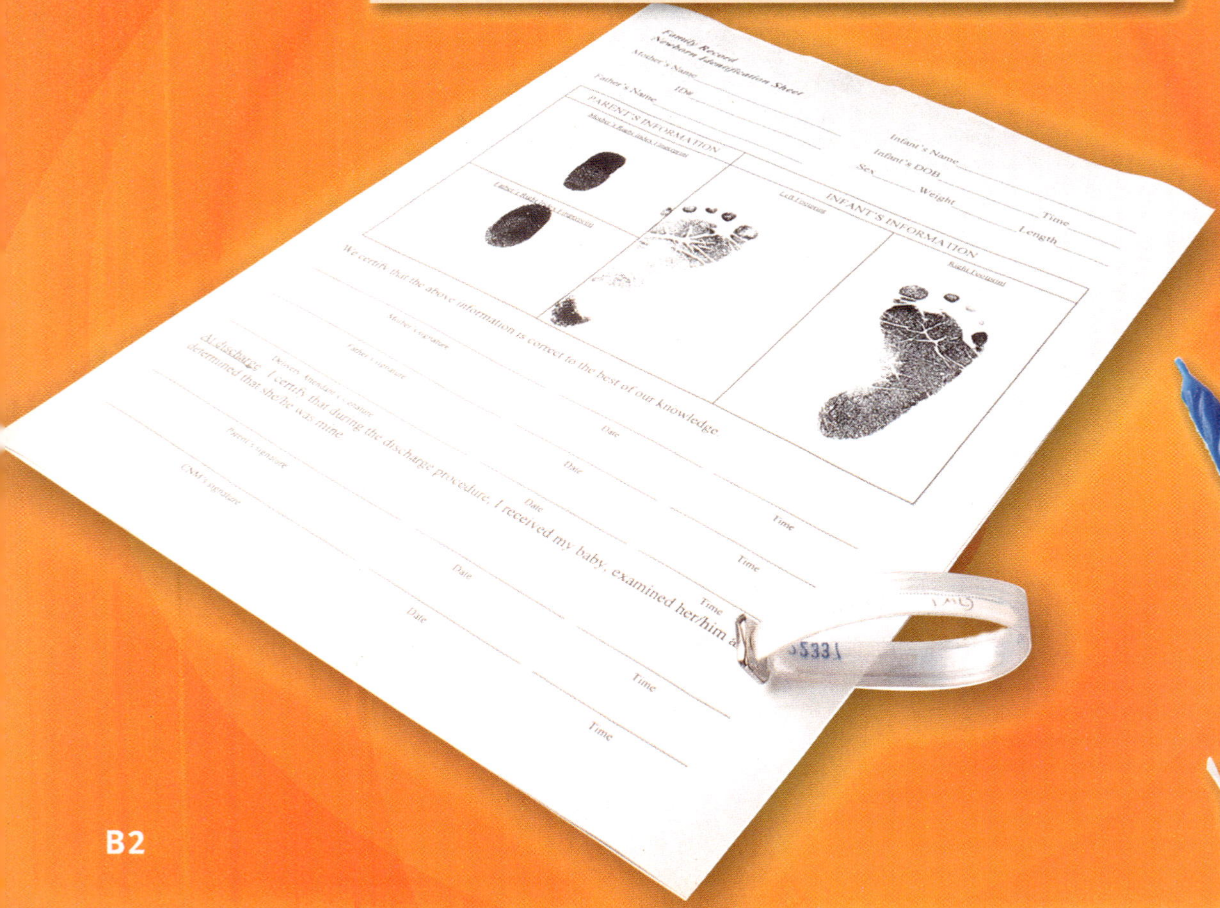

What Do You Know?

We give names to different times in a person's life. Look at the list below. See if you can match the age span to the correct time of life.

___?___	**Infancy**	**a. 18 – 70**
___?___	**Childhood**	**b. 1 – 12**
___?___	**Adolescence**	**c. birth – 1**
___?___	**Adulthood**	**d. 70 and beyond**
___?___	**Late Adulthood**	**e. 12 – 18**

Each time of life brings different changes to your body. Everyone grows and changes at different rates. What time of life are you in? Read **Growth and Development** to find out.

LESSON 1

Growing and Changing

You will learn . . .

- about the stages of the life cycle.
- about signs that show your body is changing.
- how your body grows.

Vocabulary

- **life cycle**, *B5*
- **adolescence**, *B5*
- **puberty**, *B6*
- **cell**, *B8*
- **tissue**, *B8*
- **body system**, *B8*

From the day you were born, you have been growing and changing. You have grown taller. You have become stronger. You look older. As you age, you will continue to grow and change in many exciting ways.

The Life Cycle

The stages of life from birth to death make up the **life cycle**.

Your stage of the life cycle is childhood. There are five main stages in the human life cycle. They are listed in the chart below.

Stages of the Life cycle

Stage	Description
Infancy (IN•fuhn•see)	The stage from birth to age one. You grew quickly, learned to roll over, sit up, and reach for objects.
Childhood	The stage from age 1 to age 12. During childhood you learn to manage your emotions, to solve problems, and to get along with others.
Adolescence (ad•uh•LES•uhns)	The stage from age 12 to age 18. You enter puberty and grow into an adult.
Adulthood	The longest stage of the life cycle, from age 18 to age 70. Young adults may live on their own, go to college, find a job, get married, and have a family.
Late adulthood	The stage from age 70 and beyond. Older adults may retire from their jobs, keep working, or volunteer. They experience physical and mental changes as their bodies age.

What are the five stages of the human life cycle?

Your Growing, Changing Body

As you get older, you grow taller and gain weight. Growth isn't steady, though. It often happens in *spurts*, or all of a sudden.

Growth also takes place at different rates. Your shortest classmate today might be the tallest next year. You may have some primary and some permanent teeth. Some of your classmates might have all their permanent teeth.

As your body changes, it enters puberty. **Puberty** (PYEW•buhr•tee) is the time when the body becomes able to reproduce. Girls enter puberty between ages 10 and 12. Boys enter puberty around age 12 or 13. During puberty, you will go through a big growth spurt. Your body will start to look like an adult's body.

Write About It!

Make a Time Line
Think of all of the major changes in your life. Then make a time line. Begin with the date of your birth. End with your last birthday. Mark the times when your body grew or changed the most. Include important events. Tell about each event, using complete sentences.

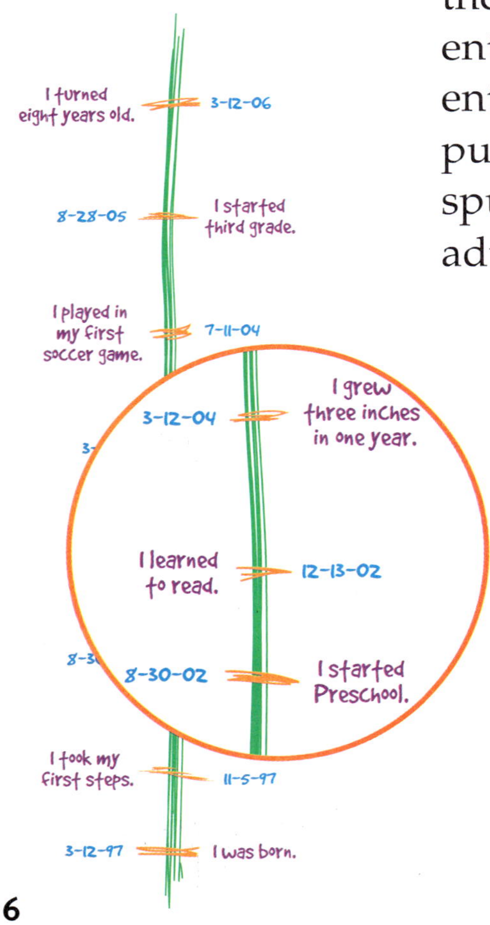

▲ As you grow, you learn new things.

Growing Up

Along with physical changes, growing up brings mental and emotional changes. It brings family and social changes, too.

Mental and Emotional As you age, you will take on more and more responsibility for yourself. You will develop many different skills and interests. You will learn how to manage your emotions.

Family and Social The activities you enjoy might change. You might have enjoyed splashing around in the pool when you were five. Now you may enjoy swimming laps or diving with friends. Today, you probably spend more time with family than with friends. When you're older, you may spend more time with friends than with family.

What is puberty?

ACTIVITY LIFE SKILLS **CRITICAL THINKING**

Access Health Facts, Products, and Services

Your friend feels terrible. She is much shorter than her classmates. What can you do to help your friend? Work with a partner. Role-play to help your friend.

1. **Identify when you might need health facts, products, and services.** What kind of information might your friend need? Make a list of your ideas.
2. **Identify where you might find health facts, products, and services.** What sources might you use?
3. **Find the health facts, products, and services you need.** Visit the library to find reliable sources or use your textbook.
4. **Evaluate the health facts, products, and services.** Was the information you found helpful? How will you use the information to help your friend?

▶ **These boys are the same age. Growth occurs at different rates for everyone.**

Your Body Systems

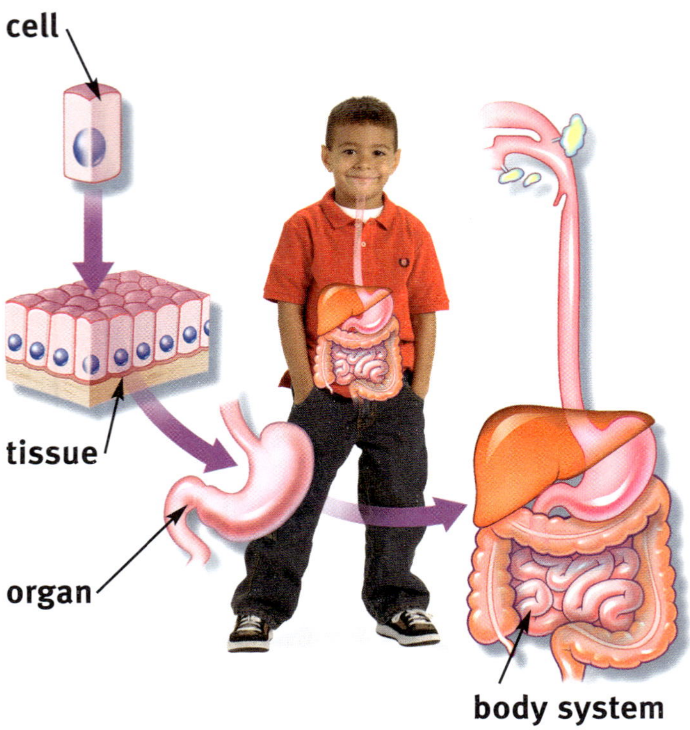

▲ **Body systems start with cells. What do tissues form?**

A **cell** is the smallest living part of the body. Some of the cells your body has are skin cells, blood cells, and bone cells. Cells are the building blocks of your body.

Some cells in your body work together. Cells that work together form a **tissue** (TI•shew). Your muscles are made up of tissue. Tissue works together to form *organs*. Your heart is an organ.

A **body system** is a group of organs that work together to do a job. Your heart, blood, and blood vessels form the circulatory system.

Major Body Systems

Body System and Function	Made Up of
Skeletal—Gives body support and shape; helps the body move	Bones, including the skull, ribs, vertebrae, and femurs
Muscular—Helps the body move; works together with bones	Muscles, including arm, thigh, and calf muscles
Circulatory—Moves blood through the body	Heart, arteries, veins
Respiratory—Helps the body use the air you breathe	Nose, throat, windpipe, lungs
Digestive—Helps the body use food to make energy	Salivary glands, stomach, small intestine, large intestine
Nervous—Controls all body actions	Brain, spinal cord, nerves, sense organs (eyes, ears, nose, mouth, skin)

Caring for Body Systems

You can help your body systems grow and stay healthy.

- **Eat healthful foods.** Eat plenty of fruits, vegetables, and fiber, such as whole-wheat bread and brown rice. Limit the amount of fatty foods you eat.
- **Get plenty of physical activity.** Walk, run, bike, skate, or swim. Play sports. Physical activity helps your body systems work better.

Name two actions you can take to care for your body systems.

▶ **To run, you use five body systems: nervous, circulatory, muscular, skeletal, and respiratory.**

LESSON REVIEW

Review Concepts

1. **Tell** what the life cycle is.
2. **List** four signs that you are growing.
3. **Explain** how your body grows.

Critical Thinking

4. **Compare** Choose two body systems from the chart on page B8. Tell how they are alike and different.
5. **Analyze** Joe exercises everyday. He also eats lots of candy and fried foods. Is Joe taking care of his body systems correctly?
6. LIFE SKILLS **Access Health Facts, Products, and Services** You've been feeling pain in your bones. You haven't been injured. You don't know what's causing the pain. What can you do?

LESSON 2

Your Muscles and Bones

You will learn . . .

- about the way your muscles work.
- what your bones do.
- ways to care for muscles and bones.

Vocabulary

- **muscle**, *B11*
- **skeleton**, *B12*
- **joint**, *B13*

Your muscles help you move. Your bones support your weight and build. It takes healthy muscles and bones to allow your body to walk, run, and do cartwheels.

How Your Muscles Work

Your *muscular* (MUHS•kyuh•luhr) *system* is a body system made up of muscles. A **muscle** (MUH•suhl) is a type of tissue made of strong fibers. Muscles let your body move. Some muscles are attached to bones. They work in pairs to help you move.

Your arm muscles help you perform many tasks. They help you lift objects and carry things. Your arm muscles help you pull open a door. They help you push your chair toward the table. Your arm muscles help you throw and catch a ball.

Your thigh muscles are in your upper legs. Your thigh muscles move your leg forward and backward. This movement takes place at the hip and knee. These muscles help you walk and run.

Your calf muscles are in the back of your lower leg. Your calf muscles bend your leg at the ankle. You need strong calf muscles to play many sports. You use calf muscles when you run, walk, ski, skate, and bike.

Name three actions that your leg muscles perform.

▲ **All the muscles in your body make up your muscular system. What do muscles do?**

How Your Bones Work

Your bones form your *skeletal system*. The **skeleton** (SKE•luh•tuhn) is the framework of your bones. The skeleton supports the body. It also protects the soft tissues inside the body. Your skeleton gives your body its shape. With your muscles, your skeleton helps you move.

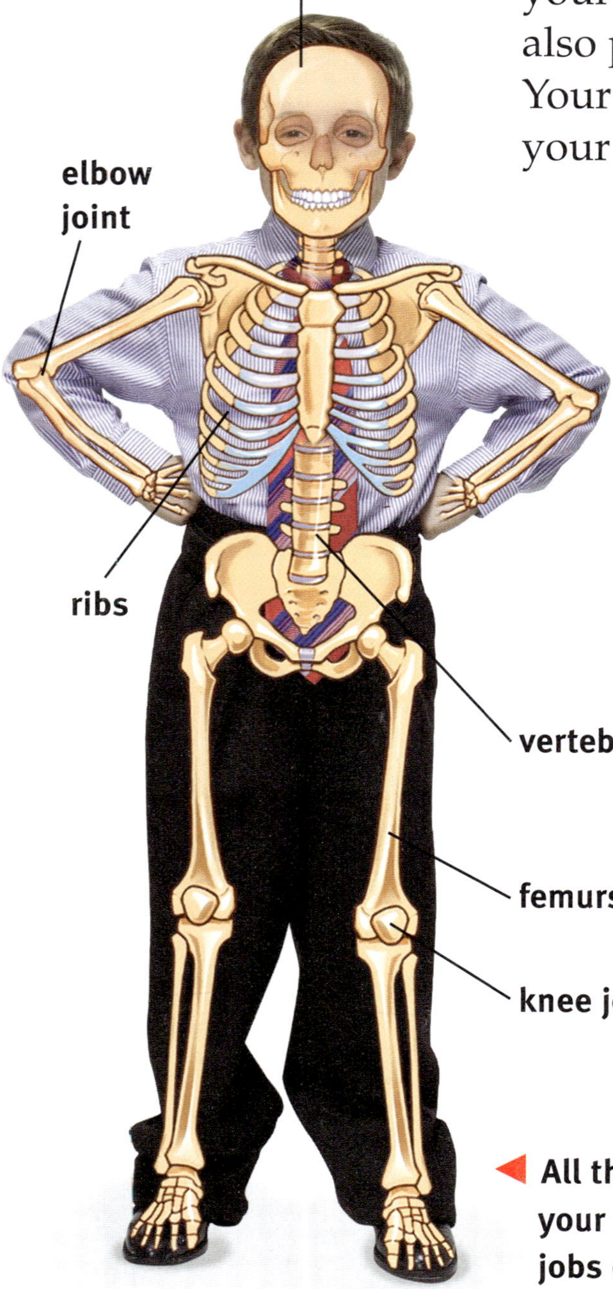

Parts of the Skeleton

There are more than 200 bones in the human body. These bones form your skeleton.

- **Your skull** is made of the bones in your head and face. There are eight different bones in your skull. Your skull covers and protects your brain.

All the bones in your body make up your skeletal system. What are the jobs of the skeletal system?

Westend61/SuperStock

- **Your ribs** are the bones that cover and protect your heart and lungs. Ribs also help support your shoulders and arms. Your ribs are arranged in pairs.
- **Your vertebrae** (VUR•tuh•bree) are the twenty-six bones that make up your spine, or backbone. Your spine supports your body and head. Your vertebrae let you bend in many directions.
- **Your femurs** (FEE•muhrz) are your thigh bones. They are the largest and strongest bones in your body.

Your Joints

A **joint** is the place where two or more bones meet. Your joints help you bend and move. The joint at your knee helps you bend your leg. The joint at your elbow helps you bend your arm. Different joints let your body move in different ways.

CAREERS

X-Ray Technician

X-ray technicians take special pictures, called *radiographs,* of different parts of the body. These pictures allow doctors to see inside a person's body. The X-ray technician in the picture is taking an X-ray of the boy's stomach. Most X-ray technicians work in hospitals. Some work in doctors' offices.

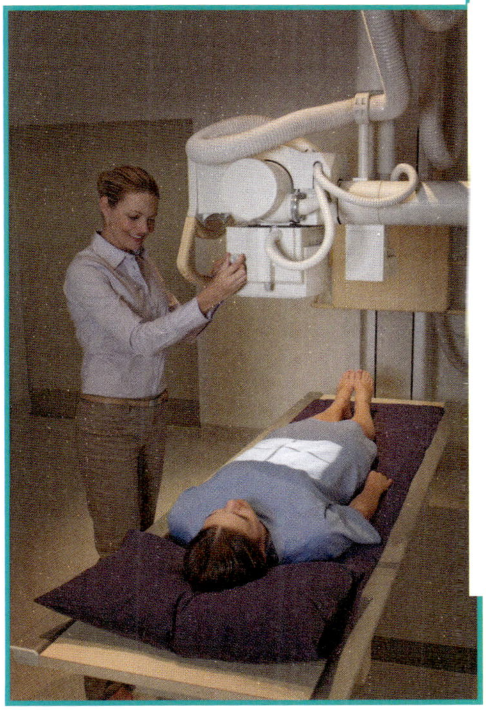

▲ **Using X rays is a kind of *technology.* Technology means "using ideas from science to do something." X rays are a kind of light that you cannot see. We cannot *see* X rays, but we can *use* them to take pictures.**

What are two jobs of the skeleton?

ACTIVITY

Physical Education LINK

Get Moving

Talk to a physical education teacher. Find out which activities can make your muscles stronger. Which ones can make your muscles more flexible? Work with a partner. Do an activity recommended by your physical education teacher in front of the class. As one partner does the activity, the other partner should explain to the class how the exercise makes muscles stronger or more flexible.

Keeping Muscles and Bones Healthy

Your body depends on strong muscles and bones. They support your body and protect your organs. Caring for your muscles and bones will make them stronger.

Caring for Your Muscles

There are a number of ways you can care for your muscles.

- **Use your muscles.** Regular physical activity makes muscles thicker and stronger. Strong muscles can help you lift, pull, push, and kick.
- **Stretch your muscles.** Stretching exercises keep your muscles flexible. *Flexibility* (flek•suh•BI•luh•tee) means being able to bend and move easily. Stretch your muscles before and after physical activity.
- **Eat foods with calcium and magnesium.** Calcium and magnesium help muscles work well. Choose milk, cheese, yogurt, green leafy vegetables, nuts, beans, cereals, breads, and seafood.

These foods are healthful for your muscles. Which of these foods contain calcium? Which ones contain magnesium?

Caring for Your Bones

There are ways to care for your bones, too.

- **Choose weight-bearing activities.** Running, walking, and dancing are good examples of *weight-bearing activities*.
- **Eat foods with calcium, phosphorus, and vitamin D.** These substances make your bones strong.
- **Sit and stand straight.** *Posture* (POS•chuhr) is the way you hold your body as you sit, stand, and move. When you have good posture, you sit and stand straight.

What are two ways to keep muscles healthy?

ACTIVITY LIFE SKILLS

CRITICAL THINKING

Set Health Goals

You want your friends to protect their muscles and bones. Work with a partner to create a poster.

1. **Write the health goal you want to set.** I will get plenty of exercise.
2. **Explain how your goal might affect your health.** How will physical activity protect muscles and bones?
3. **Describe a plan you will follow to reach your goal. Keep track of your progress.** Tell who can help you reach this goal—a teacher, a school nurse, a doctor, or a coach. Include this information on your poster. Develop a Health Behavior Contract.
4. **Evaluate how your plan worked.** What might you do if your plan fails? Include this information on your poster. Share your poster with your classmates.

LESSON REVIEW

Review Concepts

1. **Name** four actions your arm muscles perform.
2. **Identify** the parts of the skeleton.
3. **Tell** three ways you can care for your bones and muscles.

Critical Thinking

4. **Assess** Pretend that you cannot use your elbow joint. Assess how this would affect how you move.
5. LIFE SKILLS **Set Health Goals** You want to do physical activities that will get your heart working faster. What should you do?

LESSON 3

Your Heart and Lungs

You will learn . . .

- about the circulatory system.
- about the respiratory system.
- how to care for your heart and lungs.

When you exercise, you may become tired. Your face may look red. You may sweat. This is because exercise makes your heart and lungs work harder. Exercise keeps your heart and lungs strong and healthy.

Vocabulary

- **heart**, *B17*
- **oxygen**, *B18*
- **carbon dioxide**, *B18*
- **lungs**, *B18*

Design Pics/Don Hammond

The Circulatory System

The heart, blood, and blood vessels are parts of your *circulatory system*. Your **heart** is an organ that pumps blood into your blood vessels with each heartbeat. It rests between each heartbeat. Your *blood vessels* are tubes that carry blood throughout your body.

Your body has two main types of blood vessels: arteries and veins. Your *arteries* (AHR•tuh•reez) carry blood away from your heart. They move oxygen in the blood and substances to body cells.

Your *veins* bring blood back to your heart. Veins have valves. The valves keep blood from flowing backward. Muscles push against the veins to push blood to the heart.

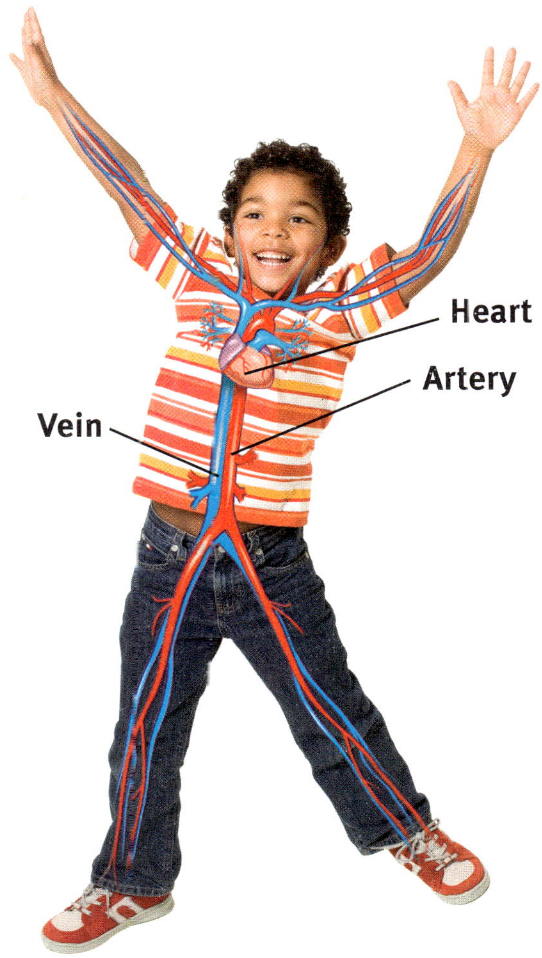

▲ The circulatory system is made up of organs that move blood throughout your body.

What are arteries?

ACTIVITY

Math LINK

Find Your Pulse

With the help of an adult, find your pulse on your wrist. Your pulse is the pressure you feel in your arteries when the aortic valve opens and closes in your heart. You can feel your pulse each time your heart beats. Count the number of pulses you feel in one minute. What number did you count? If your heart beats 75 times per minute, how many times does it beat in one hour? In one day?

The Respiratory System

Your *respiratory* (RES•puhr•uh•tor•ee) system moves oxygen into your body. It releases carbon dioxide from the body. **Oxygen** (OK•si•juhn) is a gas that you need to live. **Carbon dioxide** (KAHR•buhn•digh•OK•sighd) is a gas, too, but it is a waste product of your cells.

The respiratory system is made up of the nose, throat, windpipe, and lungs. Your nose draws air into your body. Your throat is the passageway from your mouth to your windpipe. Your windpipe is a tube that goes from your throat to your lungs.

When you breathe in, air passes through your nose and into your body. The air travels from your nose to your throat, down your windpipe, and into your lungs. Your **lungs** are organs that put oxygen into the blood.

What is carbon dioxide?

ACTIVITY

Consumer Wise

Room Air Cleaners!

People sometimes purchase a room air cleaner to make air healthier to breathe. It removes particles that irritate the lungs. Find out what other types of appliances are used to clean air in a room. You can ask your school nurse or your parents or guardian for help. Share with the class what you find out.

Your respiratory system works with your circulatory system. Blood in your arteries carries the oxygen to body cells. Blood in your veins carries carbon dioxide from body cells to the lungs. When you breathe out, or *exhale*, carbon dioxide leaves your body.

The Respiratory System

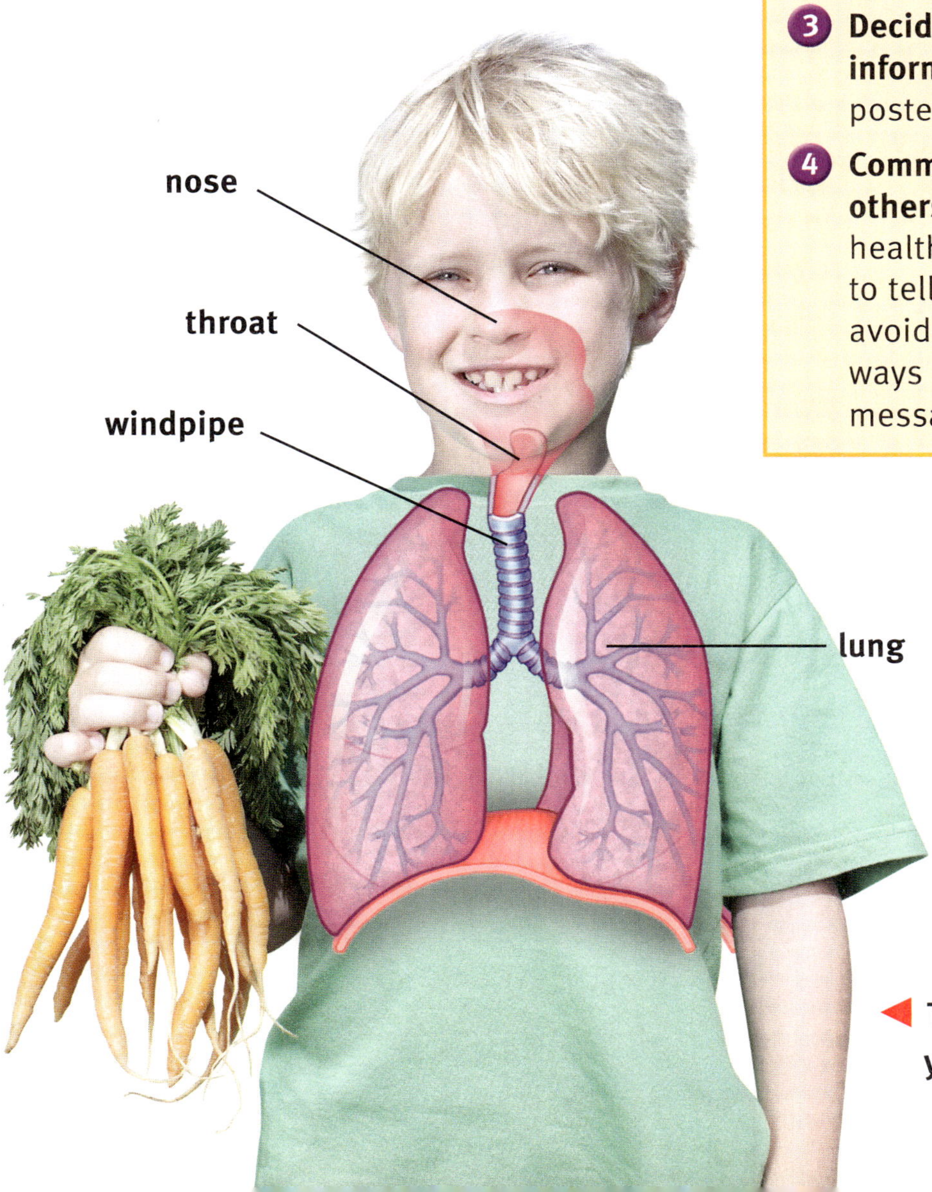

The respiratory system helps you use the air you breathe.

ACTIVITY
LIFE SKILLS
CRITICAL THINKING

Be a Health Advocate

Cigarette smoke harms your lungs. Be a health advocate and send a message against smoking. Use this activity to get your message out to others.

1. **Choose a healthful action to communicate.** Communicate why avoiding cigarette smoke is healthy for your body.
2. **Collect information about the action.** Find out how smoking harms the lungs. Record what you find.
3. **Decide how to communicate this information.** Will you make signs or posters? What else could you do?
4. **Communicate your message to others.** Follow through on your healthful action. Make a poster to tell others in your school how to avoid cigarette smoke. In what other ways might you communicate your message to others?

Caring for Your Heart and Lungs

Caring for your heart and lungs will help keep them healthy. Your heart will beat properly. You can inhale deeply, and expand your lungs fully. This will help your body get the amount of oxygen it needs.

▼ Swimming makes your heart stronger. When you have a strong heart, you can swim longer.

Your Heart

You can take action to care for your heart. Here's how.

- **Stay physically active.** Physical activity makes your heart muscle strong. It keeps fat from sticking to arteries. Walking, running, and swimming are healthful activities.
- **Eat low-fat foods.** Low-fat foods are more healthful for your arteries than high-fat foods. Substances in low-fat foods are less likely to stick to artery walls. Drink skim milk. Eat low-fat cheese and avoid fried foods.
- **Say "no" to tobacco.** Using tobacco is harmful to your heart and arteries. Tobacco products contain nicotine. Using tobacco speeds up your heartbeat. The arteries close up. Blood pressure goes up. Stay away from tobacco products.

Your Lungs

You can follow many of the same actions to care for your lungs.

- **Get plenty of physical activity.** Muscles help your lungs work. When your lungs work well, you can move without difficulty.
- **Say "no" to tobacco.** Being smoke-free keeps the respiratory system healthy.
- **Sit and stand straight.** Having good posture helps your lungs work easier. If you slouch, your lungs do not have enough space to expand.

▲ **Your posture affects how your lungs work. Sitting up straight gives your lungs room to work properly.**

Name two ways to take care of your lungs.

LESSON REVIEW

Review Concepts

1. **Name** the three parts of the circulatory system.
2. **List** the organs that make up the respiratory system.
3. **Tell** three ways to care for your heart.

Critical Thinking

4. **Explain** Why does it take both the respiratory system and the circulatory system to get oxygen to all parts of the body?
5. LIFE SKILLS **Be a Health Advocate** You want to let classmates know how they can keep their heart and lungs strong. How might you communicate this message?

LESSON 4

Your Digestive System

You will learn . . .

- how your digestive system works.
- how to care for the digestive system.

Vocabulary

- **digestion**, *B23*

The food you eat supplies your body with energy. The body system that allows this to happen is the digestive system. The digestive system breaks food down into a form the body can use.

How the Digestive System Works

Digestion (di•JES•chuhn) is the process of changing food into a form your body can use. This process takes place in your *digestive* (digh•JES•tiv) *system*. Your *salivary* (SA•luh•vayr•ee) *glands* make saliva. *Saliva* (suh•LIGH•vuh) is a liquid in your mouth that softens food. As you chew, your teeth break the softened food into small pieces. Food that is broken down into small pieces is easier to swallow. The softened food moves down a tube to your stomach.

Your stomach releases juices that break down food even more. There food is changed into a thick paste. The paste then moves into your small intestine.

Your small intestine finishes changing food into a form your body cells can use.

Your large intestine moves solid waste out of the body. Solid waste is food that cannot be digested.

What happens to food in the stomach?

ACTIVITY

Health Online

Water Works

Research the role of water in digestion. Use the e-journal writing tool to write a report.

The Digestive System

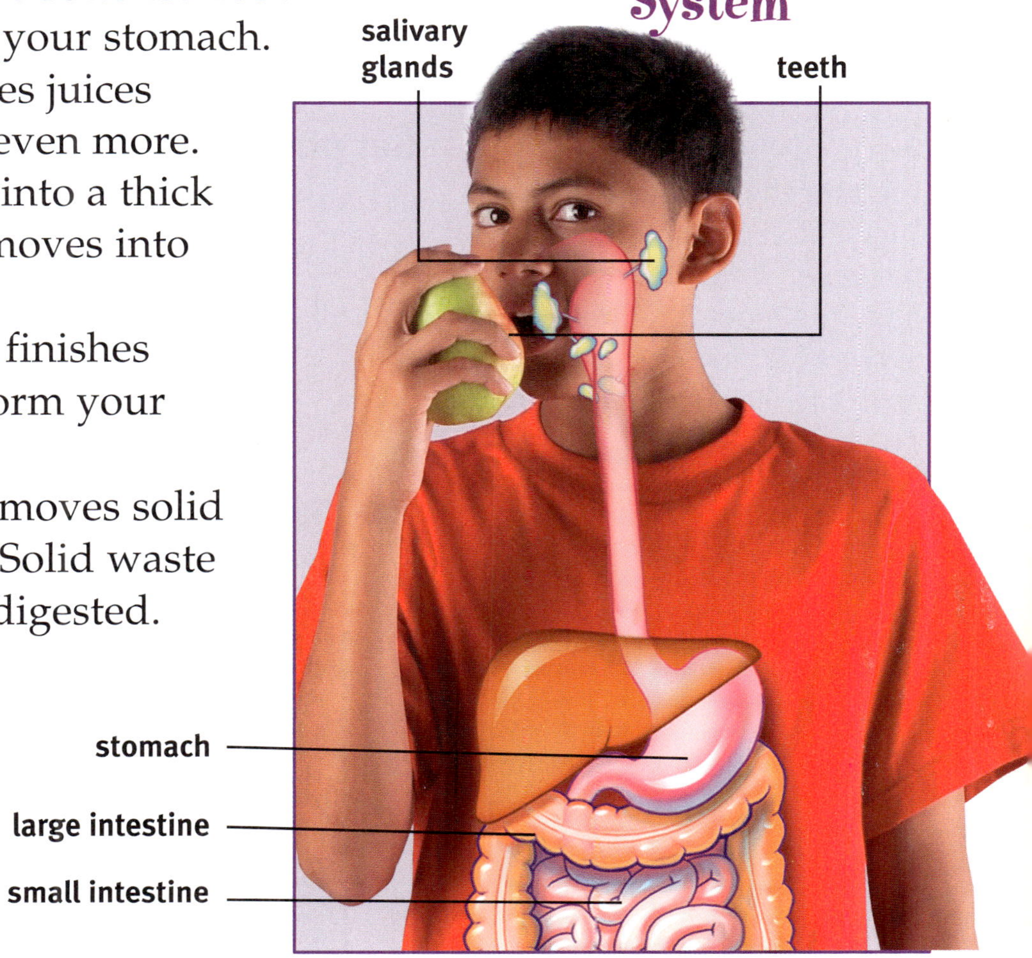

Caring for Your Digestive System

Keep your digestive system healthy by using these tips.

- **Chew your food well.** Chew and eat slowly. This gives saliva a chance to soften food.
- **Eat plenty of fruits, vegetables, and grains.** Eating foods with fiber helps your large intestine move solid waste out of your body. This keeps the large intestine clean.
- **Drink water.** Drinking water helps with digestion. You need at least six to eight glasses of water each day.
- **Get physical activity.** Regular physical activity helps muscles work. Muscles move solid waste through the large intestine and out of the body.

Which foods help your large intestine move solid waste out of the body?

ACTIVITY

On Your Own

FOR SCHOOL OR HOME

Checking Food for Fiber

Fiber is the part of plants and grains that your body cannot break down during digestion. Fruits, vegetables, and grains are foods that contain fiber. Find out how much fiber is in five foods in your house. Look for the Nutrition Facts label on the food package. Which foods have the most fiber per serving? Which ones have the least?

Nutrition Facts

Serving Size 3/4 Cup (27g)

Amount Per Serving	Cereal	With 1/2 Cup Skim Milk
Calories	90	130
Calories from Fat	10	10
	% Daily Value	
Total Fat 1g*	2%	2%
Saturated Fat 0g	0%	0%
Trans Fat 0g	0%	0%
Cholesterol 0mg	0%	0%
Sodium 190mg	8%	11%
Potassium 85mg	2%	8%
Total Carbohydrate 23g	8%	10%
Dietary Fiber 5g	20%	20%
Sugars 5g		
Protein 2g		
Vitamin A	0%	4%
Vitamin C	10%	15%
Calcium	0%	15%
Iron	2%	2%

You can find out how much fiber is in different foods by looking at the food label on the package.

ACTIVITY

LIFE SKILLS

CRITICAL THINKING

Make Responsible Decisions

After school you can choose a snack. Your choices are a candy bar or an apple. What do you do?

1. **Identify your choices. Check them out with your parent or trusted adult.** List the two choices you have.
2. **Evaluate each choice. Use the *Guidelines for Making Responsible Decisions™*.** Answer each question for each choice.
3. **Identify the responsible decision. Check this out with your parent or trusted adult.** What decision are you going to make?
4. **Evaluate your decision.** Explain why it is a responsible decision. Are you happy with your decision? Is it responsible? If not, go back to step three.

Guidelines for Making Responsible Decisions™

- **Is it healthful?**
- **Is it safe?**
- **Does it follow rules and laws?**
- **Does it show respect for myself and others?**
- **Does it follow family guidelines?**
- **Does it show good character?**

LESSON REVIEW

Review Concepts

1. **Describe** the steps involved in the digestive system.
2. **List** three actions you can take to care for your digestive system.
3. **Define** the term digestion.

Critical Thinking

4. **Connect** How might the digestive system and the circulatory system work together?
5. LIFE SKILLS **Make Responsible Decisions** You are thirsty after shooting free throws. Your friend's mom allows you to choose from her ice chest. She has soda pop and bottled water. Which should you choose?

LESSON 5

Your Senses and Nervous System

You will learn . . .

- about the five sense organs.
- what the nervous system does.
- how to care for the nervous system.

Vocabulary

- **nerve cell**, *B27*
- **brain**, *B28*
- **spinal cord**, *B29*

You get information about the world from your senses. Look around you. What do you see? Listen. What do you hear? Breathe. What do you smell? You can do all of this because of your nervous system.

The Five Sense Organs

Your body collects information using *sense organs*. **Nerve cells** carry messages from these sense organs to your brain.

All five sense organs work together. Your nerve cells work with your sense organs to carry information to your brain. Your brain interprets the information. It tells other parts of your body what to do.

Sense Organ	Description
Ears	Nerve cells inside your ears collect information about sounds and noises.
Eyes	A nerve at the back of your eyeball sends information about what objects look like to the brain.
Skin	Nerve cells deep in your skin collect information about how objects feel. They also sense cold, heat, pain, and touch.
Tongue	*Taste buds* on your tongue tell you whether something is sweet, sour, salty, or bitter.
Nose	Nerve cells inside the nose sense odors. These cells send the information to the brain.

What are the five sense organs?

ACTIVITY

Art LINK

Close Your Eyes and Draw!

Collect a piece of white paper and a pencil. While sitting at your desk, close your eyes and choose an object from the basket that your teacher is passing around the class. Keep your eyes closed and feel the object with your hands. With your eyes closed, try to draw the object that you are touching. Open your eyes. Compare the object you touched with what you drew. How did you do?

Your Nervous System

Your *nervous* (NUHR•vuhs) *system* is made up of organs. They control your body's actions. Nerve cells in these organs receive and send messages to all body parts. These messages tell your body what to do.

The Nervous System

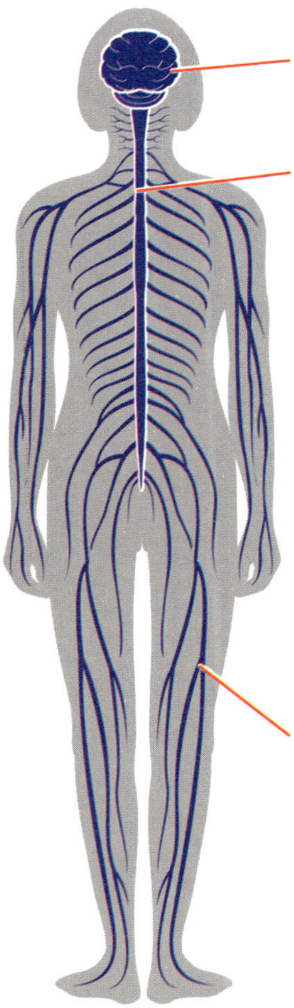

▲ **Nerves send information to your brain from different parts of your body.**

Your Brain

The **brain** is the organ that receives and sends messages to all body parts. It is made up of nerve cells. Your brain is the main organ of the nervous system.

Your brain directs everything you do. It allows your thoughts, actions, and emotions to work together. Your brain controls voluntary movement. *Voluntary movement* is action you choose to make, such as walking, talking, and eating. Your brain also controls involuntary movement. *Involuntary movement* is action that you don't control. Your heartbeat and breathing are examples of involuntary movement. They occur automatically.

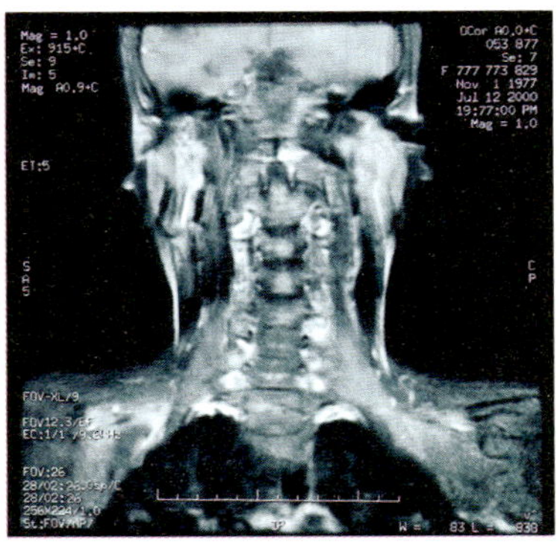

◀ **CAT scans are a kind of technology. CAT scans can give pictures of a person's brain. They can help find any problems. The earlier a problem is found, the better it can be treated and cured.**

Your Nerves

Nerves are long, thin groups of cells in your body. Some nerves carry messages from your sense organs to your brain. Suppose that a CD is playing. The nerve cells in your ears send a message to the hearing center in your brain. You hear the music.

Your Spinal Cord

The **spinal cord** is a long column of nerve cells. It sends messages to and from your brain. Your spinal cord allows messages to travel between your brain and other parts of your body. Your spinal cord is surrounded by your spine.

What are the main parts of the nervous system?

Write About It!

Your Sense Organs at Work Think of a place. It might be the beach, your bedroom, or the school cafeteria. List all the things you might see, smell, hear, taste, and feel in the place you chose. Then write a paragraph describing how this information is sent to your brain.

▶ Your nervous system helps you understand information and respond to it.

Caring for Your Nervous System

ACTIVITY BUILD Character

Analyze Your Actions

Responsibility Think about everything you do in a day. Make a list of at least 10 actions that use your nervous system. Next to each action, list a way you can take responsibility for the health of your nervous system. You need a healthy brain to do all the things you do every day. That's why it's important to take care of your nervous system.

Here's how you can take care of your nervous system.

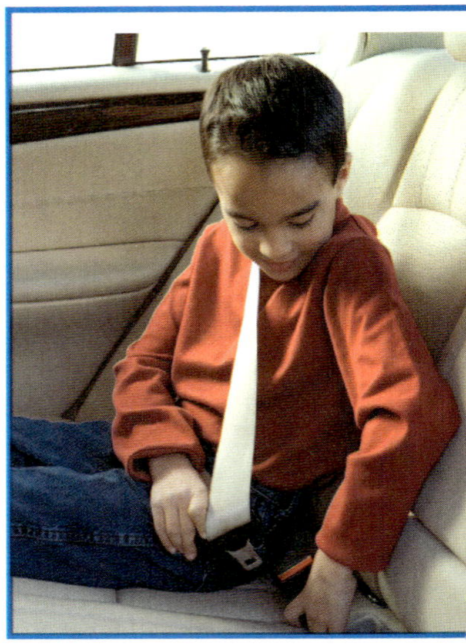

- **Wear your seat belt while riding in a car.** Wearing a seat belt protects your brain and spinal cord if you are in a car crash.
- **Say "no" to drugs.** Being drug-free helps protect nerve cells. Alcohol is a depressant drug. It slows down the action of nerve cells. Marijuana (mar•uh•WAH•nuh) is a harmful drug. It changes how your brain senses the world. Stay away from all illegal or harmful drugs.
- **Wear a helmet.** Wearing a helmet for certain physical activities protects your brain from injury. Wear a helmet when you ride your bike or scooter or skate. Many sports have rules that require players to wear a helmet. Baseball, football, and hockey are examples.

Protect Yourself from Harmful Fumes

Always protect yourself when you are around any fumes. Fumes are gasses given off by paints and glue.

- Wear a special mask when around cleaners, paints, or glues. A mask keeps you safe from breathing fumes.
- Open a window when near fumes. An open window airs out the room.

What are two ways to protect yourself from fumes?

ACTIVITY LIFE SKILLS **CRITICAL THINKING**

Use Communication Skills

Your friend lends you her scooter. You really want to take a turn. You don't have a helmet. Work with a partner. Write a skit to tell what you can do.

1. **Choose the best way to communicate.** How will you communicate with your friend?
2. **Send a clear message. Be polite.** Explain why you need a helmet. Use I-messages.
3. **Listen to each other.** Role-play using the message you wrote.
4. **Make sure you understand each other.** Tell what you heard. Then switch roles and do the activity again.

LESSON REVIEW

Review Concepts

1. **Explain** how nerve cells work with sense organs.
2. **Tell** what the nervous system does.
3. **Explain** why it's important to wear a helmet when doing certain physical activities. Give an example of an activity where you would wear a helmet.

Critical Thinking

4. **Apply** As you arrive home from school, your dog greets you. How might you react? How do your brain and nerves work when you react?
5. LIFE SKILLS **Use Communication Skills** Your older brother is driving you to school. He doesn't want to wear his safety belt. He says he doesn't need to because your school is so close by. What do you say?

Learning LIFE SKILLS

CRITICAL THINKING

Manage Stress

Problem Over the weekend, Alice broke her arm. Now she is worried. How will she get her schoolwork done. Alice may be feeling stress.

Solution Alice can manage her stress. Use the four steps on the next page to find out how.

Foldables™ To Learn Life Skills

Learn This Life Skill

Follow these four steps to help manage stress. The Foldables™ can help you.

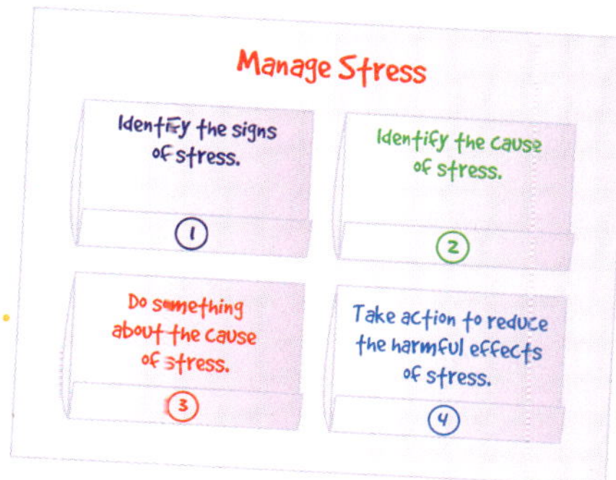

1 Identify the signs of stress.
Alice is reacting to stress. What signs of stress might she experience?

2 Identify the cause of stress.
Alice is worried about not using her arm. What do you worry about? What causes you stress?

3 Do something about the cause of stress.
Alice can talk to her parents or guardian about how she feels. What else might Alice do?

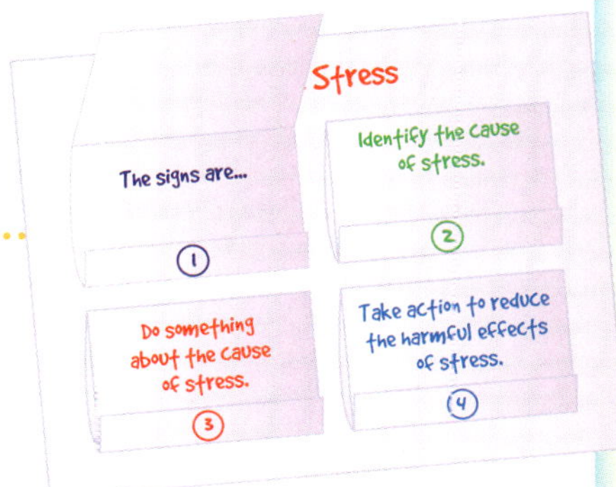

4 Take action to reduce the harmful effects of stress.
Alice can do something to take her mind off of her problem. What do you think she might do?

Practice This Life Skill

Work with a group of classmates. Think of a situation that might cause stress. Use the four steps above and write a skit about managing stress. Role-play the skit in your group.

CHAPTER 3 REVIEW

Use Vocabulary

body system, *B8*
joint, *B13*
lungs, *B18*
muscle, *B11*
spinal cord, *B29*

Choose the correct term from the list to complete each sentence.

1. The column of nerve cells that carry messages to and from the brain is called the __?__.
2. A(n) __?__ is a place where two or more bones meet.
3. Organs that put oxygen into the blood and take out carbon dioxide are __?__.
4. A(n) __?__ is a group of organs that work together to do a certain job.
5. A tissue that allows your body to move is called __?__.

Review Concepts

Answer each question in complete sentences.

6. What stages of growth do people go through after birth? Which stage are you in right now?
7. How can you take care of your muscles and bones?
8. What are the main parts of the respiratory system? What is the function of each part?
9. What are the main parts of the circulatory system? What is the function of each part?
10. List the five sense organs. What is the main job of sense organs?

Reading Comprehension

Your stomach is an organ that releases juices that break down food. Food is changed into a thick paste. The paste then moves into your small intestine.

Your small intestine is the organ that changes food into substances your body cells can use.

Your large intestine is the organ that moves solid waste, or food that cannot be digested, out of the body.

11. Which organ breaks food down into a paste?
12. Explain the function of the small intestine.
13. What happens to solid waste?

Critical Thinking/Problem Solving

Answer each question in complete sentences.

Analyze Concepts

14. Your older sister is 14. Four months ago she was 4 feet 9 inches tall. Now she is 5 feet tall. What caused her to grow so quickly?

15. After a cast is removed, how can you make a bone stronger?

16. You and your friends have been running very fast. You become tired more quickly than your friends. What can you do about it?

17. What might happen if you don't chew your food well?

18. You see the ball. You raise your foot to kick it. What organ helps you coordinate these actions?

Practice Life Skills

19. **Manage Stress** Keep a weekly log of stressful situations you experience. Describe how you felt under stress.

20. **Make Responsible Decisions** You are at a friend's house. He suggests that you jump off the top bunk of his bed onto the floor. What should you do? Use the *Guidelines for Making Responsible Decisions*™ to help you decide.

Read Graphics

The graph compares lung cancer cases with the number of cigarettes smoked per day. Use the graph to answer questions 21–23.

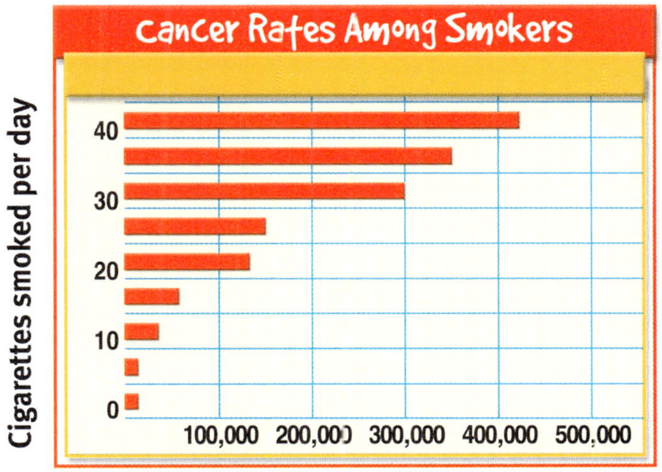

Cancer cases per 100,000 men

Source: American Cancer Society

21. How many cancer cases occur when 30 cigarettes are smoked per day?

22. How many cancer cases occur when 40 cigarettes are smoked per day?

23. How many more cancer cases occur when 40 cigarettes are smoked per day instead of 30?

CHAPTER 4

Nutrition

(tl)U.S. Fish & Wildlife Service/Ryan Hagerty, (c)Author's Image/Glow Images, (bl)lynx/iconotec/Glowimages

What Do You Know?

Are you eating enough of the foods you need each day? Do you know what foods your body needs to stay healthy? Read each statement. Answer **yes** or **no**.

___?___ Meals should include fruits and vegetables.

___?___ Eating too much sugary food is not healthful.

___?___ Food labels tell how much fat is in a serving of food.

___?___ Food gives your body energy.

Did you answer **yes** to all the questions? If so, you are off to a good start. Read **Nutrition** to find out more.

LESSON 1

Why You Need Food

You will learn . . .

- what nutrients your body needs.
- what foods are in the food groups.
- how to use MyPlate to make healthful food choices.

Vocabulary

- **energy**, *B39*
- **diet**, *B39*
- **nutrient**, *B39*
- **vitamins**, *B39*
- **MyPlate**, *B40*

To play a game of catch, you need energy. In fact, you need energy to do everything! Where do you get this energy? You get energy from food.

Scott T. Baxter/Getty Images

Your Body Needs Nutrients

Energy is the ability to do work. You get the energy your body needs from your **diet**, or the foods and drinks you digest. Foods and drinks contain nutrients. A **nutrient** (NEW•tree•uhnt) is a material in a food or drink that is used by the body. You need six kinds of nutrients. Some help you grow. Some give you energy. Others help your body work well. A healthful diet gives you all the nutrients you need.

The Six Main Nutrients	Food and Drink Sources
Carbohydrates (kahr•boh•HIGH•drayts) are your body's main energy source.	bread, pasta, rice, cereal, potatoes, fruits, fruit and vegetable juices
Proteins (PROH•teenz) are nutrients used for growth and to repair cells.	meat, fish, eggs, nuts, tofu, dry beans
Fats give you energy. They help cells store and use vitamins.	meat, dairy products, oil, butter
Vitamins (VIGH•tuh•minz) help your body use other nutrients.	vegetables, fruits, juices, meat, dairy products, cereal
Minerals keep your body working properly.	vegetables, fruits, juices, meat, dairy products
Water makes up most of your body. It helps your body use food.	drinking water, milk, fruits, vegetables

▼ Pasta is a source of carbohydrates.

▼ Chicken is a source of protein.

▼ Oil is a fat.

What is a nutrient?

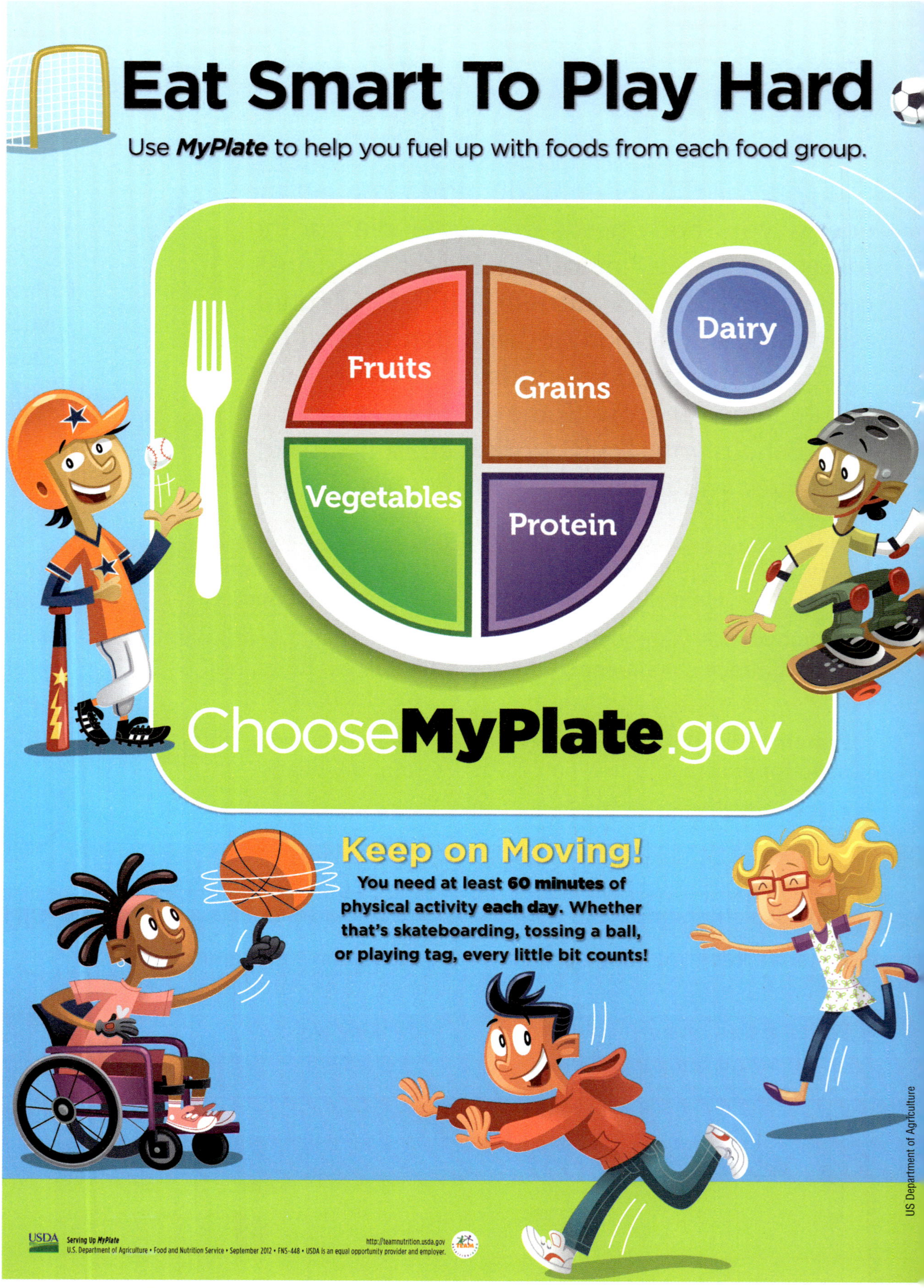

US Department of Agriculture

RUITS Fuel Up With Fruits at Meals or Snacks

ges, pears, berries, watermelon, peaches, raisins, and applesauce
out extra sugar) are just a few of the great choices.
sure your juice is 100% fruit juice.

EGETABLES Color Your Plate With Great-Tasting Veggies

eat more dark-green, red, and orange vegetables, and beans and peas.

RAINS Make at Least Half Your Grains Whole Grains

se whole-grain foods, such as whole-wheat bread, oatmeal,
e-wheat tortillas, brown rice, and light popcorn, more often.

ROTEIN Vary Your Protein Foods

ish, shellfish, beans, and peas more often. Some tasty ways include a bean
to, hummus, veggie chili, fish taco, shrimp or tofu stir-fry, or grilled salmon.

AIRY Get Your Calcium-Rich Foods

se fat-free or low-fat milk, yogurt, and cheese at meals
acks. Dairy foods contain calcium for strong
s and healthy teeth.

ow Your "Sometimes" Foods Look out for foods with added sugars or solid fats.
fill you up so that you don't have room for the foods that help you eat smart and play hard.

Using MyPlate

Don't forget to eat the correct amount from each food group every day. Here are examples.

Grains Group
You should eat 6 oz. from this food group daily. Each of these is one ounce.

1 slice of bread

1 mini bagel

1 cup ready-to-eat cereal

1/2 cup cooked cereal, rice, or pasta

5 whole wheat crackers

Vegetable Group You should eat 1 to 1.5 cups from this food group daily.

Raw, leafy vegetables

Cooked or chopped raw vegetables

Vegetable juice (100%)

Fruit Group You should eat 1 to 1.5 cups from this food group daily.

Apples, bananas, or oranges

Chopped, cooked, canned, or mixed fruit

Fruit juice (100%)

Dairy Group You should eat 2.5 cups from this food group daily. Use low-fat or fat-free products.

Milk or yogurt

Natural cheese, like cheddar

Processed cheese

Proteins Group

You should eat 4 oz. of these foods daily.

Cooked lean meat, fish, or poultry

Eggs

Peanut butter

How many cups of fruit should you eat?

ACTIVITY LIFE SKILLS

CRITICAL THINKING

Access Health Facts, Products, and Services

Your doctor tells you to get plenty of Vitamin C in your diet. How do you know which foods and drinks have Vitamin C?

1. **Identify when you might need health facts, products, and services.** Find health facts about foods and drinks that contain Vitamin C.
2. **Identify where you might find health facts, products, and services.** Find three sources with health facts about food, drinks, and Vitamin C.
3. **Find the health facts, products, and services you need.** Use all three sources. Write down foods and drinks with Vitamin C.
4. **Evaluate the health facts, products, and services.** Ask your doctor if your facts are correct.

LESSON REVIEW

Review Concepts

1. **Name** the six kinds of nutrients your body needs.
2. **Name** three foods in each of the five food groups.
3. **Tell** how you can use MyPlate to have a healthful diet.

Critical Thinking

4. **Analyze** Why is it important to eat fruits and vegetables?
5. LIFE SKILLS **Access Health Facts, Products, and Services** You want to learn more about vitamins. What sources could you use?

LESSON 2

The Dietary Guidelines

You will learn . . .

- about the Dietary Guidelines.
- how to follow the Dietary Guidelines.
- how to use the Dietary Guidelines to choose snacks that are healthful.

Vocabulary

- **Dietary Guidelines**, *B45*
- **snack**, *B48*

Do you know about the Dietary Guidelines? Do you know how to use them? If so, you can choose healthful meals and snacks.

Donald Iain Smith/Getty Images

The Dietary Guidelines

The **Dietary Guidelines** are suggested goals to help you stay healthy. These guidelines help you stay healthy and live longer. They were developed by the U.S. government. There are nine Dietary Guidelines. They list the kinds of foods you should eat. They also list the kinds of foods you should limit or avoid. The Dietary Guidelines explain why you should stay at a healthful weight. They also explain why you should be physically active every day.

What is the goal of the Dietary Guidelines?

ACTIVITY

Consumer Wise

Compare Fruit Drinks

Gather a label from a fruit juice and a fruit drink. Read each label. Compare. Which has more sugar? Which product is more healthful?

CAREERS

Cafeteria Worker

The cafeteria workers at your school help you get the nutrients you need. They may plan meals using MyPlate and the Dietary Guidelines. Cafeteria workers prepare or cook food. They also serve you food.

How to Follow the Dietary Guidelines

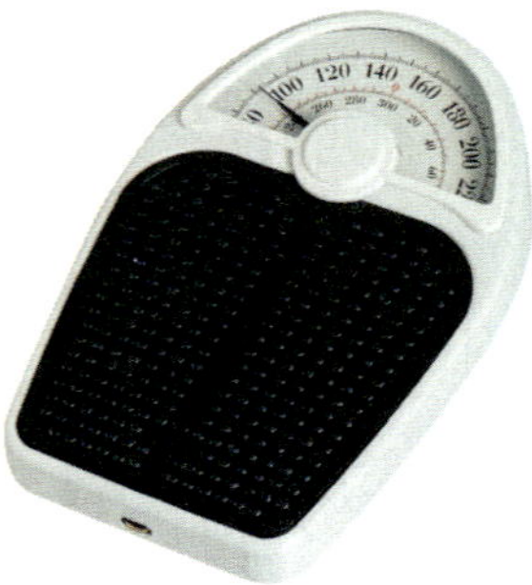

Follow the Dietary Guidelines to stay healthy and live longer. You'll have more energy, too.

1. **Get the calories you need for how active you are.** A *calorie* is a unit used to measure the energy produced by food in the body. Let MyPlate guide your food choices to get the calories you need.

2. **Stay at a healthful weight.** Being overweight is a health problem. It increases the risk of heart disease, diabetes, and some cancers. Your doctor can tell you how much you should weigh. Eat the right amount of food and get plenty of exercise to maintain a healthful weight.

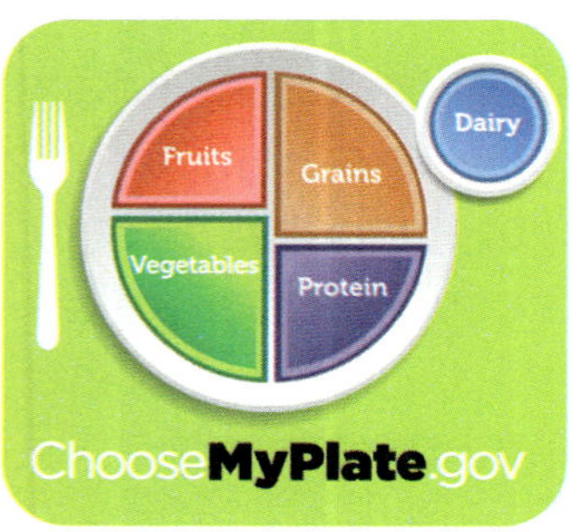

3. **Be physically active each day.** Physical activity helps you stay at a healthful weight. It also makes your heart, lungs, and muscles strong. It helps you manage stress. You need about 60 minutes of physical activity every day.

4. **Get the nutrients you need from healthful foods.** Eating more fruits, vegetables, whole grains and low-fat or fat-free milk helps you get the vitamins, minerals and fiber that you need.

5. **Choose foods that do not have too much fat.** Limit the meat you eat. When you eat meat, cut the fat off it. Limit pastries, such as donuts, cookies, and cake. Nuts, vegetables, and fish are healthful choices for fats.

6. **Choose foods and drinks that do not have too much sugar.** Candy, cake, and many soft drinks are high in sugar. Eating too much of these foods can make you gain weight. Sugar also can cause tooth decay.

7. **Choose and prepare foods without much salt.** Eating too much salt increases your risk of some diseases. You need less than 1/4 teaspoon of salt each day. Many packaged or canned foods have salt added to them. Try not to add salt to your food. Also eat few salty snacks, such as chips and pretzels.

8. **Keep food safe to eat.** Germs can get into food and make you ill. Wash your hands before you eat or prepare food. Keep dairy foods such as milk cold.

9. **Do not drink alcohol.** Alcohol is a drug. It can harm body organs. It is against the law for people your age to drink alcohol.

Why is one goal of the Dietary Guidelines to stay at a healthful weight?

Write About It!

Stay Healthy, Live Longer Review the Dietary Guidelines listed on pages B46 and B47. Write a summary explaining how the guidelines can help you stay healthy and live longer. Include a plan telling how you can make the Dietary Guidelines a part of your life.

How to Choose Healthful Snacks

A **snack** is food eaten between meals. You can use the Dietary Guidelines to help you choose healthful snacks. Here are some tips.

- **Choose snacks to help you get the right amounts from the food groups.** Use MyPlate to help you.
- **Choose snacks made from grains.** Whole wheat crackers and cereal are healthful snack choices.
- **Choose fruits and vegetables for snacks.** An apple or orange and carrot sticks help you get the servings needed from the fruit and vegetable food group.
- **Limit snacks that are high in total fat.** These food items include cake, candy, French fries, and hamburgers.
- **Limit snacks that have added sugar.** Avoid soft drinks, fruit drinks with sugar, and candy.
- **Limit snacks with salt.** Choose unsalted pretzels, crackers, and peanuts. Do not add salt to popcorn.

▲ Fruit makes a healthful snack.

What is a snack?

ACTIVITY

LIFE SKILLS

CRITICAL THINKING

Set Health Goals

Are you ready to follow the Dietary Guidelines? Set a goal.

1. **Write the health goal you want to set.** I will eat healthful snacks.
2. **Explain how your goal might affect your health.** Read the Dietary Guidelines again. Suppose you choose snacks that follow the guidelines. List three ways this helps your health.
3. **Describe a plan you will follow to reach your goal. Keep track of your progress.** Make a health behavior contract. Your contract should include a chart with columns for each day of the week. For one week, write down the snacks you eat each day. Tell who can help you—a parent or guardian, a teacher, a school nurse, a school nutritionist.
4. **Evaluate how your plan worked.** After one week, check your chart. Did all the snacks follow the Dietary Guidelines? Did you reach your goal? How could you make your plan better?

LESSON REVIEW

Review Concepts

1. **Tell** what the Dietary Guidelines are.
2. **List** two of the nine Dietary Guidelines.
3. **Identify** two ways to choose healthful snacks.

Critical Thinking

4. **Synthesize** How are the Dietary Guidelines and MyPlate related?
5. LIFE SKILLS **Practice Healthful Behaviors** You want to limit the salt you eat. How will you use the four steps to practice this healthful behavior?
6. LIFE SKILLS **Set Health Goals** You want to get plenty of physical activity every day. Write a health behavior contract to help you reach this goal.

LESSON 3

Choosing Healthful Meals and Snacks

You will learn . . .

- what influences your food choices.
- what facts are found on food labels.
- how to plan a menu for a healthful meal.

Vocabulary

- **ad**, *B51*
- **food label**, *B52*
- **ingredient**, *B52*

You make food choices every day. You might help write your family's shopping list. What influences your food choices? Do you read food labels? Do you clip coupons? Do you compare prices?

Influences on Food Choices

Suppose you are choosing healthful meals and snacks. Why do you choose some foods and drinks and not others? There are probably different reasons for your choices.

- **Personal Preference** You eat many foods because you like the way they taste or smell.
- **Family and Friends** You may eat certain foods because your parents or guardian serve them at meals. Your family's culture may influence food choices. You may have tried some foods served by your friend's parents or guardian when eating at their home.
- **Availability** You may eat a food that is grown in your area. You may choose foods that many people in your community eat.
- **Food Ads** Food companies want you to choose the foods they sell. An **ad** is a notice that tells people about a product. Companies use ads to influence your food choices.
- **Dietary Guidelines** Using the Dietary Guidelines to choose healthful foods and drinks will keep you healthy and help you live longer.

What is an ad?

BUILD Character ACTIVITY

Make Responsible Food Choices

Responsibility You are responsible for your health. You show this responsibility when you check out food ads. Choose a food ad from a magazine. Look at the ad. Does the food ad tell you how the food fits with the Dietary Guidelines? Explain.

Family and friends influence your food choices.

Food Labels

Check food labels and prices to help you make wise choices.

ACTIVITY

Math LINK

Compare Packages

The food store is having a sale on canned peas. A 6-ounce can of peas is $0.69. A 15-ounce can of peas is on sale for 1 cent more. How many more ounces of peas do you get for the extra penny you pay?

Read Food Labels

Most packaged foods have food labels. A **food label** lists the ingredients in the food. It also lists nutrition facts. An **ingredient** (in•GREE•dee•uhnt) is something that goes into making a food or drink. The label lists a food's ingredients in order. The first one listed makes up most of the product. The last one listed is the least ingredient in the product.

Nutrition Facts
Serving Size 1 cup (228g)
Serving Per Container 2

Amount Per Serving	
Calories 250	Calories from Fat 110
	% Daily Value*
Total Fat 12g	18%
Saturated Fat 3g	15%
Trans Fat 1.5g	
Cholesterol 30mg	10%
Sodium 470mg	20%
Total Carbohydrate 31g	10%
Dietary Fiber 0g	0%
Sugars 5g	
Protein 5g	
Vitamin A	4%
Vitamin C	2%
Calcium	20%
Iron	4%

* Percent Daily Values are based on a 2,000 calorie diet. Your Daily Values may be higher or lower depending on your calorie needs:

	Calories:	2,000	2,500
Total Fat	Less than	65g	80g
Sat Fat	Less than	20g	25g
Cholesterol	Less than	300mg	300mg
Sodium	Less than	2,400 mg	2,400mg
Total Carbohydrate		300g	375g
Dietary Fiber		25g	30g

▲ **This food label for macaroni and cheese gives nutrition information.**

The Nutrition Facts panel is another part of a food label. This panel tells about the nutritional value of the food. It tells you how much of the food makes up a serving. It also tells how many nutrients are in one serving.

Compare Prices

You can also compare prices to make your choice. If the food labels are similar, but one brand costs less, it might be a better choice.

What is an ingredient?

ACTIVITY LIFE SKILLS CRITICAL THINKING

Make Responsible Decisions

You want a snack. You can choose between two boxes of cereal. One is sugar coated. The other is not.

1. **Identify your choices. Check them out with your parent or trusted adult.** You could choose the sugary cereal or the other cereal.
2. **Evaluate each choice. Use the *Guidelines for Making Responsible Decisions*™.** Answer each question using both of your choices.
3. **Identify the responsible decision. Check this out with your parent or trusted adult.** Make your choice.
4. **Evaluate your decision.** Explain why your choice is responsible.

Guidelines for Making Responsible Decisions™

- **Is it healthful?**
- **Is it safe?**
- **Does it follow rules and laws?**
- **Does it show respect for myself and others?**
- **Does it follow family guidelines?**
- **Does it show good character?**

Planning a Menu

> **MAKE a Difference**
>
> Students helped feed Florida families whose homes were found unsafe. These families had to move to new places. After moving they had little money left to buy food. The students at Saint Charles Borromeo school took action. They formed the *Kids Care Club* and collected food. Then they delivered the food to the families.

A food plan is also called a *menu*. You order from a menu at a restaurant. You may help your parents or guardian plan a family menu for one day.

How to Plan a Menu

You can plan a menu for healthful meals. Here's how.

- **Review the Dietary Guidelines.**
  The Dietary Guidelines are suggested goals for eating.

When planning a menu, also include healthful snacks that may be eaten during the day. Foods from many countries around the world can be healthful snacks.

- **Follow MyPlate.** Choose foods and drinks from each food group. This will help you get the right amounts from each food group. Include three meals and snacks in your menu.
- **Think about food safety.** Choose foods that do not need to be kept cold if you are going to a picnic.

You may choose healthful foods from many countries around the world. Eggplant dishes and chick peas are tasty examples. So are yucca and plantains. Tacos and pita sandwiches are favorite treats as well.

ACTIVITY

Health Online

Same Food, Different Price

Research and report on a food product that is sold at two different prices. Compare ingredients. Use the e-Journal writing tool to report your findings.

What is a menu?

LESSON REVIEW

Review Concepts

1. **Name** three influences on your food choices.
2. **Explain** what kinds of information are found on food labels.
3. **Tell** how you can plan a menu for a healthful meal.
4. **Explain** why you might compare prices when shopping for a food item.

Critical Thinking

5. **Compare** The first three ingredients listed for one cereal are sugar, oats, and corn. The first three ingredients listed for another are oats, corn, and sugar. Which cereal has more sugar? How do you know?
6. LIFE SKILLS **Make Responsible Decisions** It's time for a snack. A friend offers to trade his candy bar for your apple. What should you do?

Learning **LIFE SKILLS**

CRITICAL THINKING

Analyze What Influences Your Health

Problem You are at a fast-food restaurant with a friend's family. Everything looks good. What influences your food choices?

Solution You can analyze what influences your food choices. Use the four steps on the next page to help you.

Foldables™ To Learn Life Skills

Learn This Life Skill

Follow these four steps to help analyze what influences your food choices. The Foldables™ can help you.

1 Identify people and things that can influence your health.

Your friends influence your food choices. Who else might influence you?

2 Evaluate how these people and things can affect your health.

How do these influences affect your health? Record your ideas.

3 Choose healthful influences.

Decide which influences help you to make healthful food choices at the restaurant.

4 Protect yourself against harmful influences.

Be aware of influences that may result in unhealthful food choices at the restaurant.

Practice This Life Skill

Work with a group. Write a skit about what influences your food choices. Use the steps above.

LESSON 4

Food Safety

You will learn . . .

- how food can be kept safe.
- how kitchen safety and table manners keep you safe.

Vocabulary

- **spoiled**, *B59*

You grab a jug of milk out of the refrigerator. Look at the date. This date can help tell you if the milk is still safe to drink. Reading dates is just one way to make sure food is safe to eat.

Keep Foods Safe

Foods and drinks can become **spoiled**. This means they are unsafe to eat. A food or drink spoils because germs grow on or in it. These germs can make you ill.

Food Safety Tips

Prepare Food

- Wash your hands before you prepare food. Also wash them before you eat.
- Wash your hands after you touch raw meat or eggs.
- Wash fresh fruits and vegetables before you eat or cook them.
- Wash a cutting board that has been used to cut raw meat. Then you can use it to cut other foods.
- Check dates on foods and drinks before using them.

Cook Food

- Cooking kills germs. Cook meat well. A meat thermometer can be used to see if it has cooked enough.
- Do not eat foods with raw eggs. The eggs can contain germs. Unbaked cookie dough and cake batter may have raw eggs in them.

Clean Up

- Wrap leftovers quickly. Then put them into the refrigerator. Wrapping keeps germs off foods. Cold makes germs grow more slowly.
- Clean counters and cooking surfaces with soap and warm water. Dry with a paper towel.

What makes food spoil?

ACTIVITY

On Your Own

FOR SCHOOL OR HOME

Check Food Packages

Many foods and drinks have dates on their packages. These "use by" or expiration dates tell you when a food may not be safe to eat. Check the dates on foods and drinks in your refrigerator. Make a chart and keep track of all of the foods and their expiration dates. Decorate your chart and put it on your refrigerator.

Kitchen Safety and Table Manners

Do you like to help make meals? Do you enjoy eating with others? Stay safe. Follow kitchen safety rules when making meals. Use table manners when eating.

Kitchen Safety

When you work in the kitchen, be careful. Here are some safety rules.

- **Don't use a kitchen appliance unless your parent or guardian says it's okay.** A kitchen appliance (uh•PLY•uhns) is a small machine used for preparing food. Use it only with an adult.
- **Do not use an appliance near water or when you are wet.** You could get shocked.
- **Do not touch hot stove tops.** Hot stove tops cause burns.
- **Use pot holders.** This protects your hands from heat.
- **Never stick an object into a toaster.** You could get shocked.
- **Use a knife carefully.** Use it only with an adult.
- **Handle glasses and sharp utensils carefully.** You can cut yourself on broken glass or sharp blades.

▼ **Be careful when you wash glasses and sharp utensils.**

Use Table Manners

Table manners are polite ways to eat. Using table manners shows respect for others.

- Place your napkin in your lap to protect your clothing.
- Wait until everyone is served before you start eating.
- Chew with your mouth closed.
- Do not talk with your mouth full.

Why is it important to use table manners?

ACTIVITY
LIFE SKILLS
CRITICAL THINKING

Practice Healthful Behaviors

You are going to a friend's house for dinner. How can you use table manners?

1. **Learn about a healthful behavior.** Identify the table manners you need to know before you go to dinner. Which will you use?
2. **Practice the behavior.** Practice the table manners when you visit your friend. Which table manners make eating safer and more pleasant?
3. **Ask for help if you need it.** If you have a question about table manners, ask.
4. **Make the behavior a habit.** Practice these manners at dinner. Write a paragraph about the manners you used.

LESSON REVIEW

Review Concepts

1. **Name** three ways to keep germs out of food.
2. **Describe** how to clean counters and cooking surfaces after you have prepared food.
3. **Explain** why you should not talk with your mouth full.

Critical Thinking

4. **Infer** Why should you use a kitchen appliance only when a parent or guardian is watching?
5. LIFE SKILLS **Practice Healthful Behaviors** Suppose you are helping to wash dishes. Why should you check carefully before you reach into the dishwater?

CHAPTER 4 REVIEW

Use Vocabulary

- **diet,** *B39*
- **Dietary Guidelines,** *B45*
- **MyPlate,** *B40*
- **food label,** *B52*
- **nutrients,** *B39*
- **spoiled,** *B59*
- **vitamins,** *B39*

Choose the correct term from the list to complete each sentence.

1. The guide that tells you the amounts from each food group your body needs every day is _?_.
2. The materials in food that are used by the body are _?_.
3. The food and drink usually eaten by a person is called a(n) _?_.
4. The _?_ are suggested goals to help you stay healthy and live longer.
5. When foods and drinks become _?_, they are unsafe to eat.
6. Nutrients that help your body use other nutrients are called _?_.
7. A(n) _?_ lists the ingredients and nutrition facts of a packaged food item.

Review Concepts

Answer each question in complete sentences.

8. What nutrient is the body's main energy source?
9. How can you use MyPlate to plan a healthful diet?
10. What are three influences on your food choices?
11. Why is it unsafe to eat uncooked cookie dough?
12. What are table manners?

Reading Comprehension

Answer each question in complete sentences.

The Vegetable Group is the green food group in MyPlate. You need to eat 1 to 1.5 cups from this group daily. Foods in the Vegetable Group give you vitamins, minerals, and carbohydrates.

13. What is the green food group in MyPlate?
14. How many cups of vegetables should you eat each day?
15. Why is it healthful to eat different vegetables for each meal?

Critical Thinking/Problem Solving

Answer each question in complete sentences.

Analyze Concepts

16. What can you learn from reading a food label?

17. Explain why you should eat some whole grain foods each day.

18. Give three reasons you should not drink alcohol.

19. List the kinds of nutrients the body needs. Why is each nutrient important?

20. Why is it important to wrap and refrigerate leftover foods promptly?

Practice Life Skills

21. **Analyze What Influences Your Health** Some snacks that are sold in stores do not follow Dietary Guidelines. What influences may make you want to buy healthful snacks?

22. **Make Responsible Decisions** You are spending the night at a friend's house. Your friend offers you a fruit drink. Your parent or guardian does not want you to drink fruit drinks because they are high in sugar. What should you do? Use the *Guidelines for Making Responsible Decisions*™ to help you decide.

Read Graphics

The chart shows the amounts from each food group a person should eat daily. How much did John eat? Use the chart to answer questions 23–24.

Food Group	Recommended Amount	Number of servings John ate in one day
Protein	4 oz.	6 oz.
Dairy	2.5 cups	3 cups
Fruits	1 to 1.5 cups	1 1/2 cups
Vegetables	1 to 1.5 cups	1/2 cup
Grains	6 oz.	6 oz.

23. From which food groups did John get the recommended amount of food?

24. From which food group did John have an extra amount?

UNIT B ACTIVITIES AND PROJECTS

Effective Communication

Make a Poster

Make a poster that lists safety rules for preparing food. Draw pictures to go with the list. Hang the poster in your kitchen.

Self-Directed Learning

Make a Timeline

Visit the library or interview an adult. Learn about the changes that happen to people as they grow from birth to your age. When do they learn to walk? When do they start losing baby teeth? Make a timeline showing when changes happen.

Critical Thinking and Problem Solving

Plan a Menu

Some people don't eat meat. They are vegetarians. Plan a healthful menu that doesn't have meat. Include breakfast, lunch, and dinner.

Responsible Citizenship

Donate Food

Some families do not have the healthful foods they need. Ask your parent or guardian if you can donate food to a food pantry to help these families. Choose foods and drinks that are healthful.

UNIT C

Personal Health and Safety

CHAPTER 5

Personal Health and Physical Activity

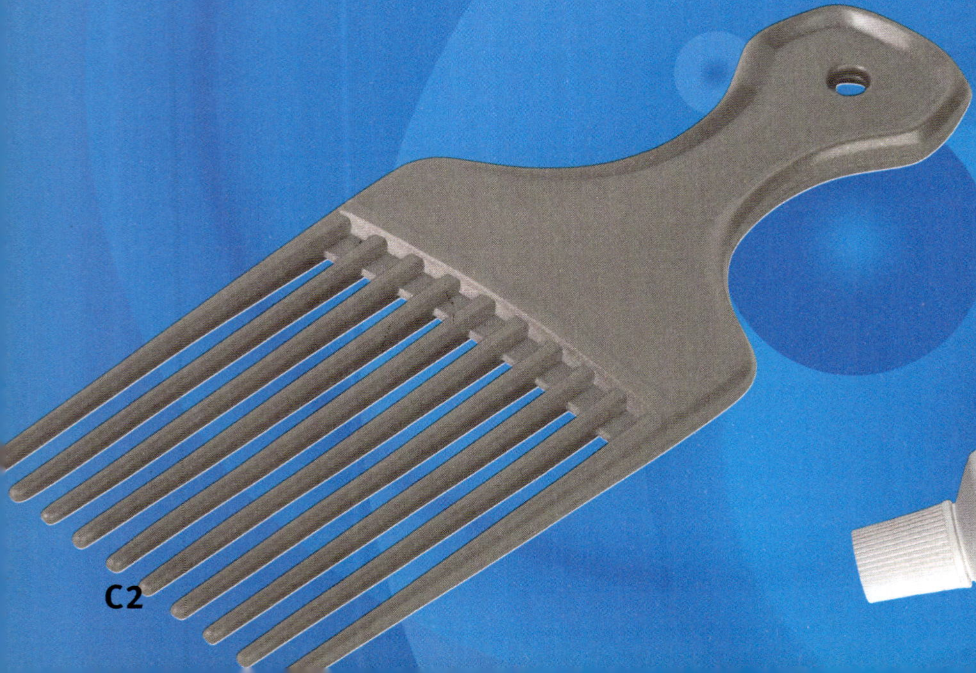

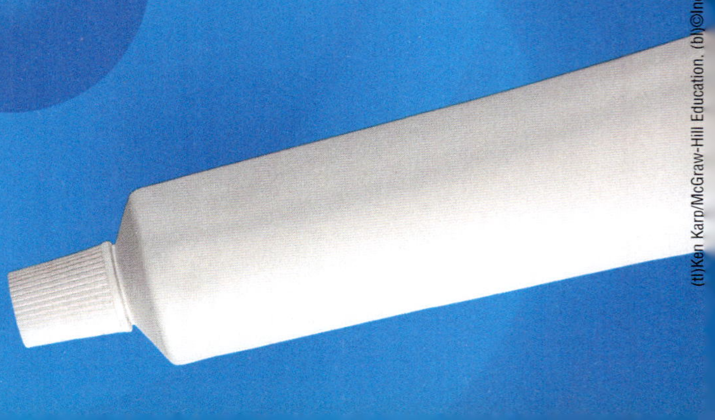

(tl)Ken Karp/McGraw-Hill Education, (b)©Ingram Publishing/Fotosearch, (bc)©Comstock/Alamy, (cr)©Pixtal/SuperStock

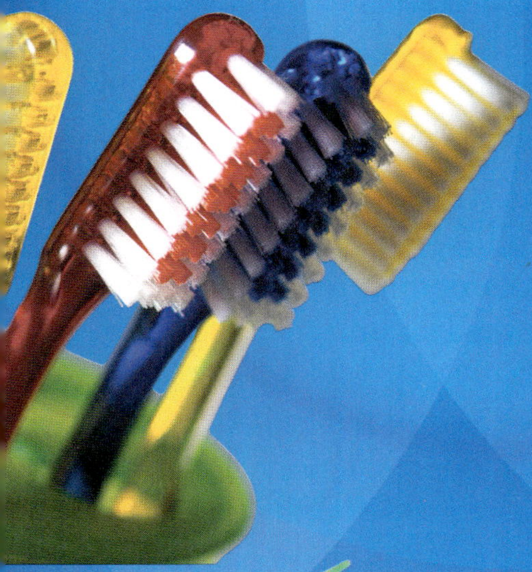

What Do You Know?

Do you practice healthful habits? Answer **yes** or **no** to the following health habits.

_?_ I floss and brush my teeth every day.
_?_ I use proper lighting to protect my eyes when I read or watch television.
_?_ I wash my hair with shampoo.
_?_ I stretch before I exercise.
_?_ I wear a helmet when I ride my bike.

Did you answer **yes** to all five statements? If so, you are practicing healthful habits. Read **Personal Health and Physical Activity** to learn about these and other health habits.

LESSON 1

Checkups and Dental Health

You will learn . . .

- why you need medical checkups.
- why you need to keep your teeth healthy.
- how to floss and brush your teeth.

Vocabulary

- **checkup**, *C5*
- **dental plaque**, *C6*
- **cavity**, *C6*
- **dental floss**, *C7*

You can take action to stay healthy. You can wash your face everyday. You can get the proper amount of sleep each night. Having regular checkups can help you stay healthy, too.

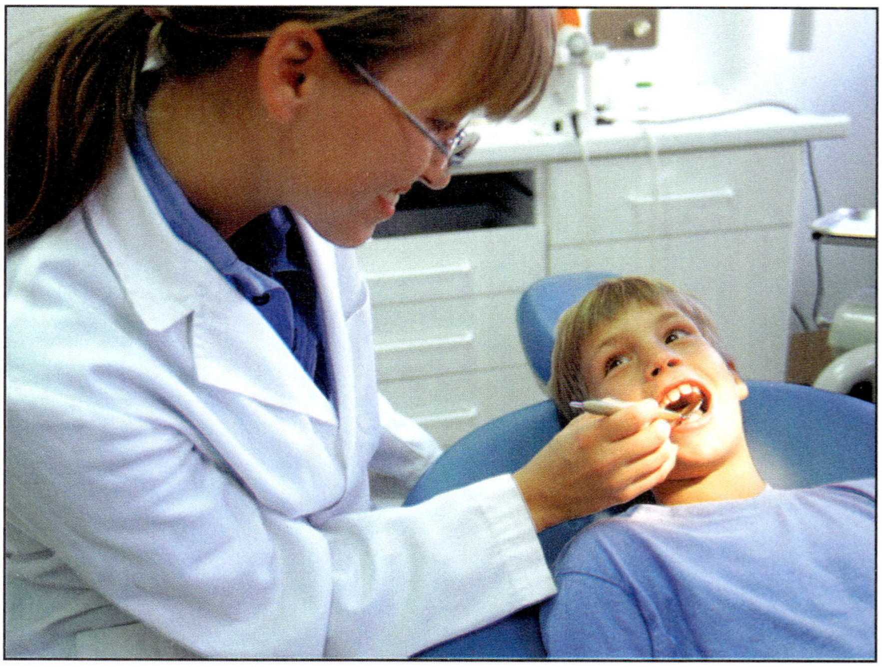

Medical Checkups

A doctor's job has many parts. Doctors help people who are ill. Sometimes they give ill people medicine. They might give advice, too.

Doctors also help people stay well. They give checkups. A **checkup** is a medical exam to make sure you are healthy. During a checkup, the doctor measures your height and weight. The doctor checks your eyes, ears, nose, and throat. He or she checks your lungs and heart. Checkups help you and your doctor notice changes in your health.

Personal Health Record Your doctor keeps a health record on you. A personal health record tells facts about your health history. It includes changes in your weight and height and the times you were ill. You and your parents or guardian can keep a personal health record at home.

Going to the Doctor You might feel stress seeing the doctor. *Stress* is the response to any demand on your mind or body. Young people feel stress when they are nervous or excited. Talk to your parents or guardian about your feelings. Talking may help you feel better.

CAREERS

Pediatrician

A pediatrician is a doctor who takes care of babies and children. Their growth and development is very important. A pediatrician looks for changes in a child's health. He or she can help a child make a health plan.

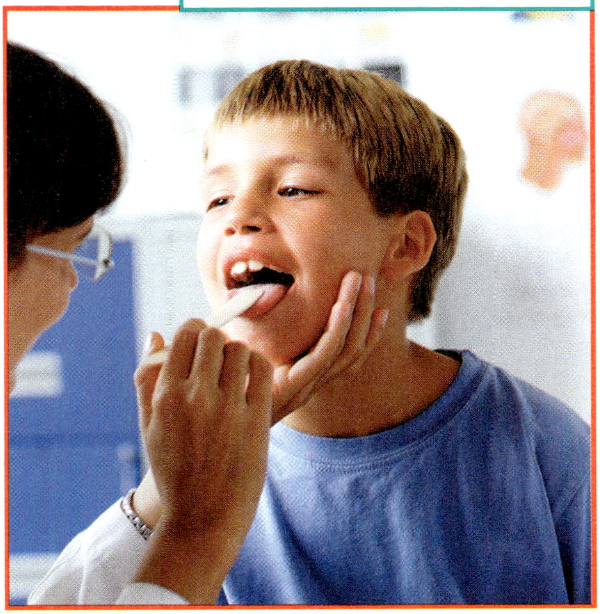

▲ A doctor who takes care of children is called a pediatrician.

Why is it important to have regular medical checkups?

Healthy Teeth

Do You Know

Dentistry developed in Egypt around 3700 BC. People first started going to the dentist as we do today during the 1800s.

There are three ways healthy teeth help you. They give you a nice smile. They help you speak clearly. They bite and chew food. As you chew pieces of food can stick to your teeth. This food mixes with germs in your mouth. Then a sticky material called **dental plaque** forms on your teeth. If dental plaque stays on your teeth, a cavity may form. A **cavity** is a hole in a tooth.

A dental health plan can help keep your teeth and gums healthy.

Parts of a Tooth

- **Crown** The part above the gums
- **Enamel** A hard, white material that covers the crown
- **Dentin** Bony layer under the enamel
- **Pulp** The soft, inner part that has nerves and blood vessels
- **Root** Holds the tooth in the jawbone

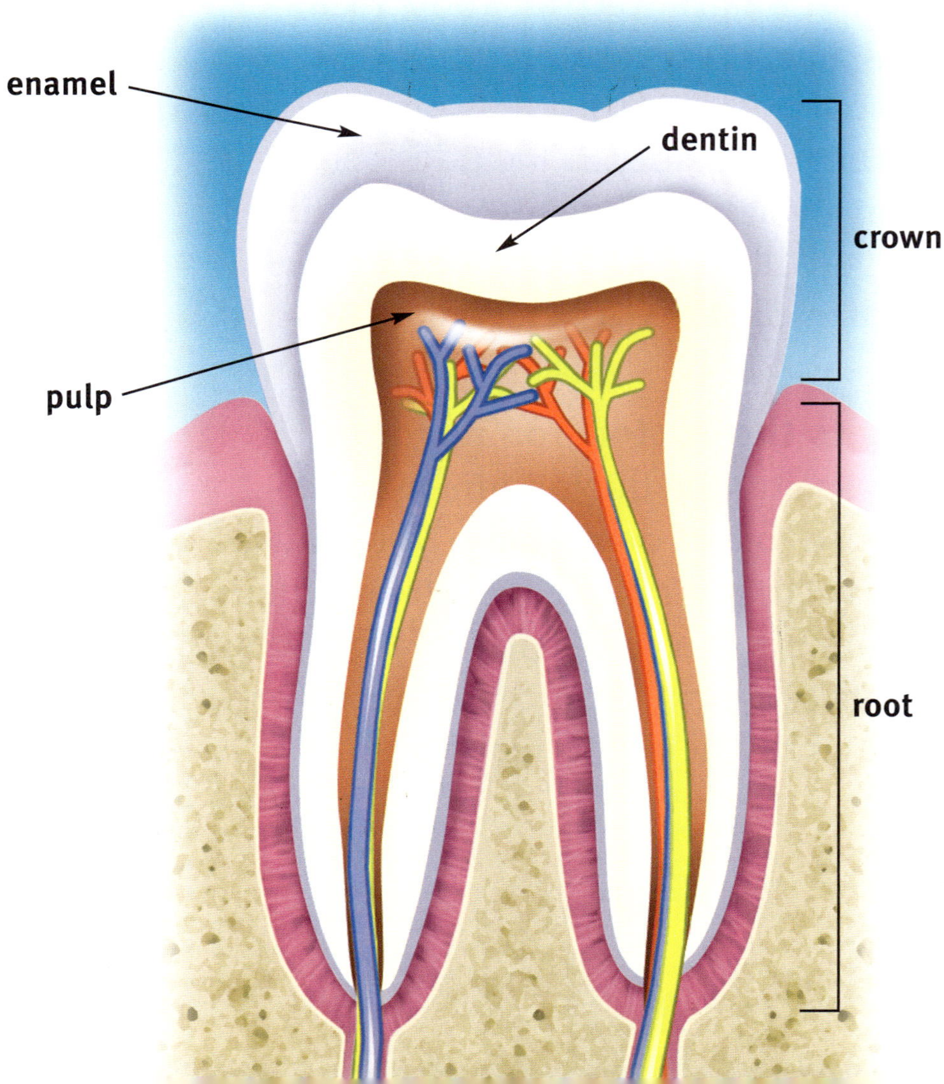

Caring for Your Teeth

You can care for your teeth. Here's how.

- **Use a toothbrush and toothpaste.** Brush your teeth at least two times a day. Brushing helps remove food from your teeth and gums.

- **Floss your teeth once a day.** Use dental floss. **Dental floss** is a string-like material. It is used to remove food and dental plaque stuck between teeth and under gums.

- **Eat healthful foods.** Foods made with milk help keep your teeth strong. Foods that have vitamin C in them keep gums healthy. These foods include oranges, other fruits, and tomatoes.

- **Avoid foods or beverages that have a lot of sugar in them.** Sugar increases your chance of having cavities.

- **Have a dental checkup every six months.** Your dentist will check your teeth and gums and help you make a dental health plan.

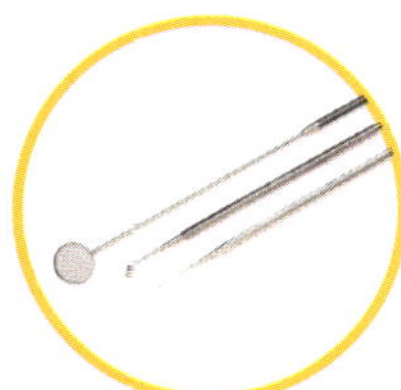

- **Wear a mouthguard when you play contact sports like football or hockey.** A mouthguard protects your teeth from injury.

Why should you limit foods and drinks with sugar?

Flossing Your Teeth

Floss before you brush. Follow these steps.

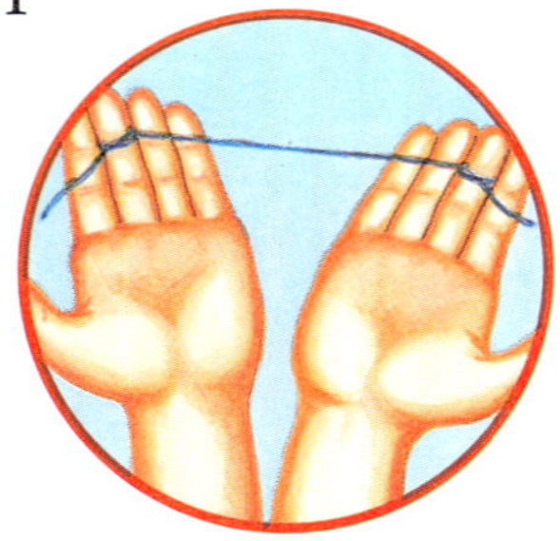

1 Start with 18 inches of floss. Wrap most of the floss around your middle fingers. Hold the floss tightly between your thumb and forefingers.

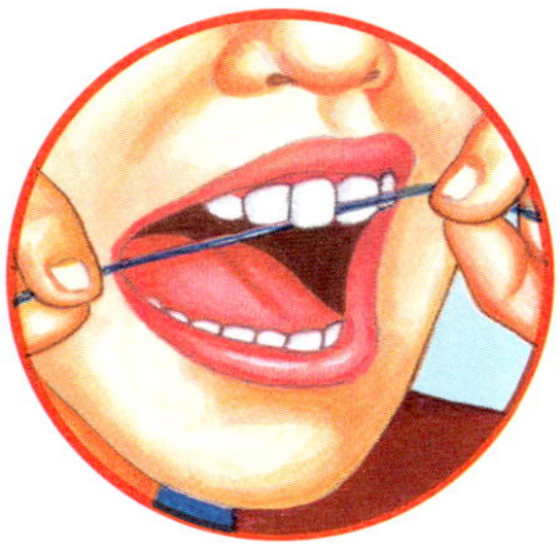

2 Gently slide the floss between two teeth and into the space between the gum and tooth. Curve the floss around the tooth. Move the floss away from the gum using up and down motions.

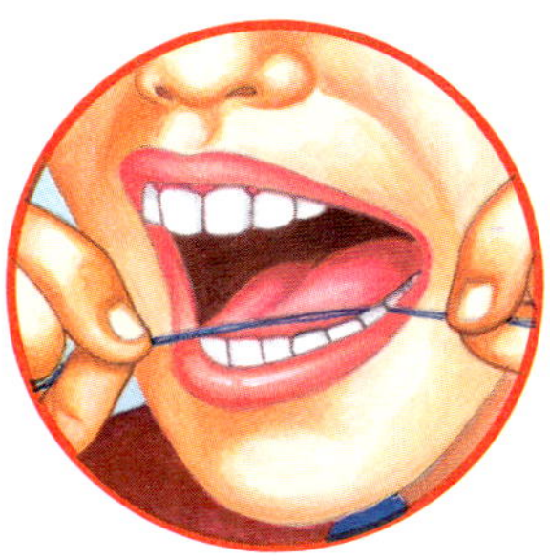

3 Repeat these steps with each tooth.

ACTIVITY LIFE SKILLS **CRITICAL THINKING**

Make Responsible Decisions

Role-play: Tyrone asks Matthew to play football without his mouthguard.

1. **Identify your choices. Check them out with your parent or trusted adult.** What are Matthew's choices?
2. **Evaluate each choice. Use the *Guidelines for Making Responsible Decisions™*.**

Guidelines for Making Responsible Decisions™

- **Is it healthful?**
- **Is it safe?**
- **Does it follow rules and laws?**
- **Does it show respect for myself and others?**
- **Does it follow family guidelines?**
- **Does it show good character?**

3. **Identify the responsible decision. Check this out with a parent or trusted adult.** What should Matthew do?
4. **Evaluate your decision.** Why is Matthew's decision responsible?

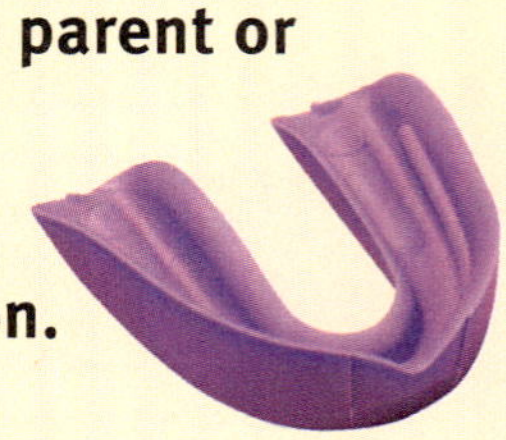

Brushing Your Teeth

Take at least two minutes twice a day to follow these steps.

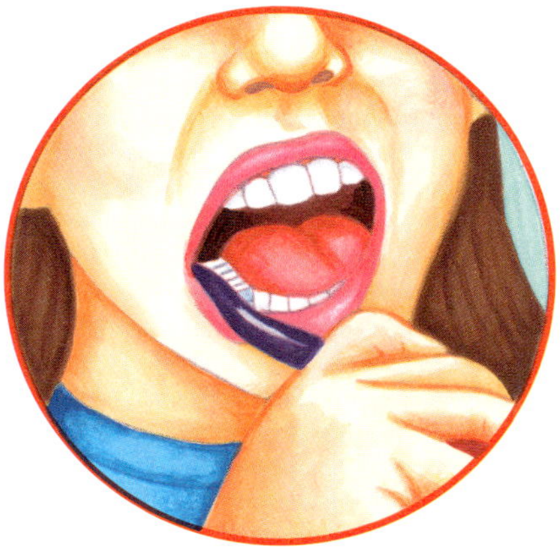

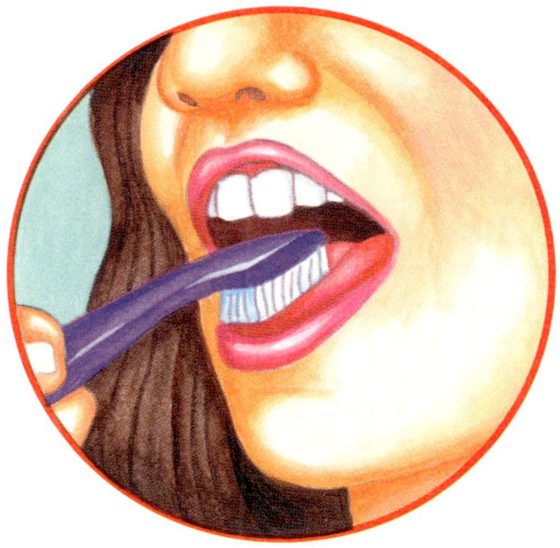

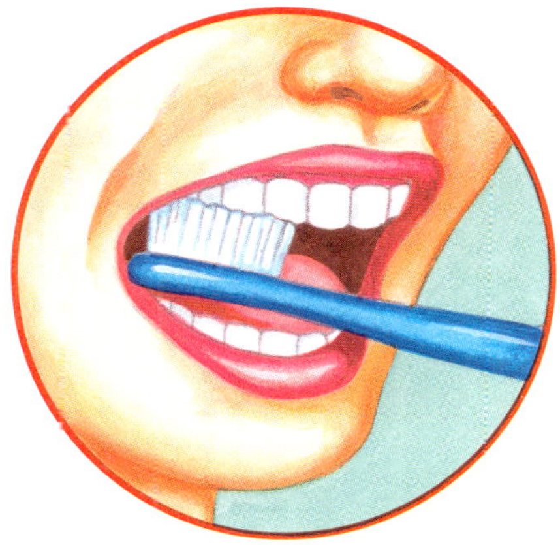

1 Place your toothbrush against your teeth. Tilt the bristles against the gums. Move up and down in a circle over the outer surface of your teeth.

2 Brush your tongue. This helps get rid of germs from your mouth. It also helps keep your breath fresh.

3 Brush the inside surface of your teeth. Then brush the chewing surfaces of your back teeth.

LESSON REVIEW

Review Concepts

1. **Explain** what is included in a medical checkup.
2. **Summarize** the correct way to brush your teeth.

Critical Thinking

3. **Analyze** why your food and drink choices are included in a dental health plan.
4. LIFE SKILLS **Make Responsible Decisions** You are with your family at a fast-food restaurant. It is your turn to order. You have set a health goal to take care of your teeth. Use the *Guidelines for Making Responsible Decisions*™ to help you choose a beverage.

LESSON 2

Your Eyes and Ears

You will learn . . .

- how to keep your eyes healthy.
- how to keep your ears healthy.
- how to protect your hearing.

Your eyes and ears are two of your sense organs. They take in facts about the world around you. You can take care of your eyes and ears. You can protect your vision and hearing. You can get regular checkups.

Vocabulary

- **vision**, *C11*
- **ear canal**, *C12*
- **eardrum**, *C12*

Getty Images/Fuse

Your Eyes

You rely on your sense of sight, or **vision**, in many ways. Your eyes help you stay safe. For example, you look up and down the street to check for cars before crossing. Your eyes also help you communicate. You show respect when you look at someone you are talking to.

Take Care of Your Eyes If your eyes itch or burn, tell your parents or guardian right away. You also could tell your teacher or the school nurse. Itching and burning can be signs of an eye problem. You may have an infection such as pinkeye. *Pinkeye* is an infection of the membrane covering the inner eyelid. The eyelids become inflamed and vision is sometimes blurred. Eye infections should be treated quickly. Ways to protect your vision are listed at the right.

Why is it important to take care of your eyes?

◀ **Some visually impaired people read using a system of writing called *Braille.* In Braille, dots are put together in certain ways to make the letters of the alphabet.**

Protect Vision

- Use proper lighting when you read, watch television, or work on the computer.
- Wear sunglasses that protect your eyes from the sun's harmful rays. Never stare at the sun.
- Wear safety glasses when you play sports.
- Don't rub your eyes if something gets in them. Ask an adult to help you.
- Be careful with sharp objects. You could injure an eye.
- Tell an adult if you get hit in the eye or if your eyes itch or burn.
- Get regular checkups.

Your Ears

Sound is all around you. Sometimes you react to the sounds you hear. You wake up when the alarm clock goes off in the morning. You may jump when a car horn blows. You might hurry to breakfast when you hear eggs sizzling in a pan. Nerves in your ears send information about sound to your brain.

Sound enters your ear through the ear canal. The **ear canal** leads from your outer ear to your eardrum. The sound travels to your eardrum. Your **eardrum** is a thin tissue that is stretched like the top of a drum inside your ear. Sound causes your eardrum to move back and forth. Nerves in your ear pick up the motion and send messages to your brain. Your brain interprets the sounds you hear. If your eardrum is injured, your sense of hearing may be reduced or lost.

Write About It!

Write a Poem Keeping your ears healthy is an important part of a health plan. Write a poem giving reasons you value your sense of hearing. Include ways to protect your hearing.

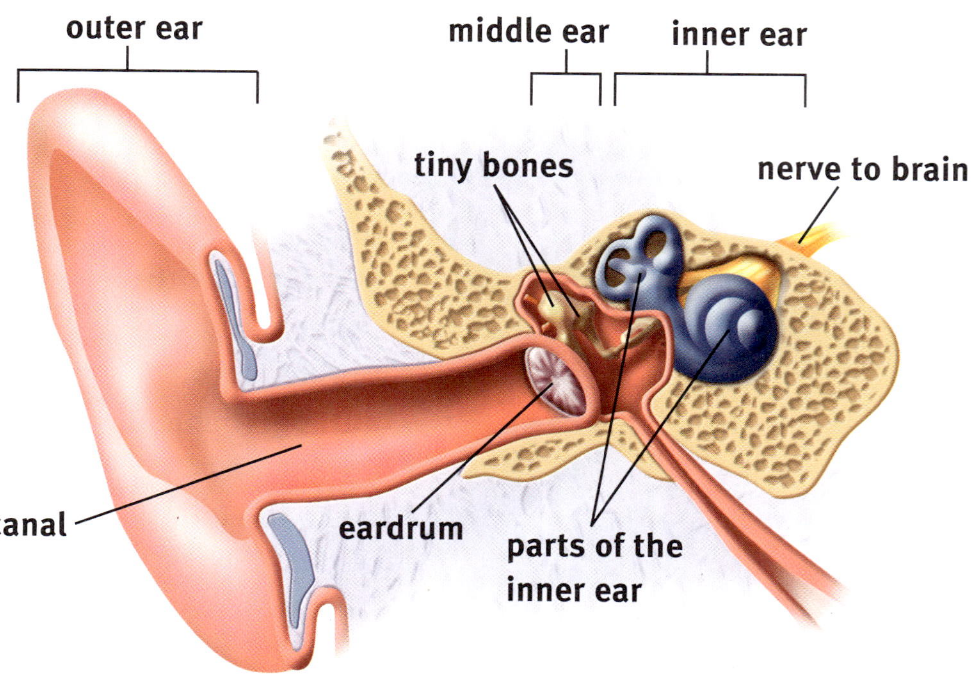

▲ **When the eardrum moves back and forth, the movement travels through parts of the middle and inner ear.**

Protect Hearing

Here's how you can protect your hearing.

- **Tell an adult if you have an earache or your ear itches.** Your ear may be infected. It should be treated quickly.
- **Never stick objects, including cotton swabs, into your ears.** You could injure your eardrum.
- **Avoid loud sounds.** Loud sounds can cause hearing loss. Play music and television at regular volumes. Wear headphones or earplugs when around loud sounds.
- **Get regular checkups.**

What should you do if you have an earache?

ACTIVITY LIFE SKILLS CRITICAL THINKING

Access Health Facts, Products, and Services

1. **Identify when you might need health facts, products, and services.** You want to make a pamphlet to put in your classroom listing sounds that can hurt hearing. The pamphlet also will tell how to protect your ears from that sound.
2. **Identify where you might find health facts, products, and services.** With a partner, list sources of this information. For example you could use books in your school library as sources.
3. **Find the health facts, products, and services you need.** Find facts from two different sources to use in your pamphlet.
4. **Evaluate the health facts, products, and services.** With your partner, decide which sounds are most important for your pamphlet.

LESSON REVIEW

Review Concepts

1. **List** three ways to protect your vision.
2. **Explain** why you should avoid loud sounds.
3. **Tell** three ways to help protect your hearing.

Critical Thinking

4. **Summarize** what you would do if you got a blow to your eye.
5. LIFE SKILLS **Access Health Facts, Products, and Services** Find out more information about sunglasses. Which types of sunglasses offer the best protection?

LESSON 3

Good Grooming

You will learn . . .

- why grooming your skin, hair, and nails is important.
- why getting enough rest is important to good health.

Vocabulary

- **skin**, *C15*
- **sunscreen**, *C15*
- **grooming**, *C16*
- **health-care product**, *C16*

Washing your face gives you a healthy appearance. Taking care of your skin, hair, and nails helps you look your best. These behaviors keep you well-grooned.

Your Skin

Your **skin** is an organ that covers your body. It prevents germs from entering your body. Your skin keeps sunlight from harming the inside of your body.

It is important to keep your skin healthy. Eating a balanced diet helps. Foods that have vitamin A are good for your skin. Green vegetables, orange vegetables, eggs, and milk are sources of vitamin A.

The sun's harmful rays can damage the skin. They can cause skin cancer. Sunburn happens when you are in the sun without protection. Using sunscreen can help prevent your skin from burning. **Sunscreen** is a lotion that is put on the body to block the sun's harmful rays. To protect your skin, use a sunscreen that has a Sun Protection Factor (SPF) of at least 15.

Why should you use sunscreen?

Write About It!

Write a Skit Suppose you are going to the beach. Write a skit in which you get a sunburn. Explain why wearing sunscreen is important to protect your skin. Act out the skit with a small group.

Using sunscreen with an SPF of at least 15 every time you are in the sun can help protect you from skin cancer later in life.

Grooming

Grooming is taking care of your body and appearance. Regular grooming helps keep your skin healthy. Washing your skin with soap and water every day removes dirt, oil, and germs. Washing your hands regularly removes germs. It helps keep germs from getting into your eyes and mouth.

A **health-care product** is something that is used to promote health. There are health-care products for grooming. Shampoo, sunscreen, and hand cream are health-care products. Soap, toothpaste, and floss are health-care products you might use every day.

Some products may cause side effects. A *side effect* is an unwanted feeling or illness after using a substance. Itching skin or a rash are side effects. Tell your parent or guardian if you have side effects after you use a product.

Using Health-Care Products

- Ask your parents or guardian before you use a new grooming product. They can help you decide if the product is right for you.
- Read the label. Look for warnings and read them closely.
- Follow the directions for proper use of the grooming product.
- Tell your parents or guardian if you have side effects.

A nurse can check your hair for lice.

Your Hair

Wash your hair with shampoo to remove dirt and oil. Use shampoo that is made for your hair type. Conditioner coats hair and makes it easier to comb.

Sharing hats, brushes, or combs with others can spread head lice. Head lice, shown on page C16, are tiny insects that lay eggs in the hair. To treat head lice, use a special shampoo that kills the lice and eggs.

Your Nails

Your nails protect the ends of your fingers and toes. Trimming your nails is part of good grooming. Nails that are trimmed and smooth are less likely to be damaged. Clean nails by washing them with soap and water. Do not bite your nails or pick at the skin around the nails.

What is an important part of grooming skin, hair, and nails?

Do You Know

The longer the finger, the faster the nail grows. The nail on the middle finger grows the fastest. Fingernails also grow faster than toenails. Toenails are thicker than fingernails.

▲ Use a nail brush to clean under nails.

▲ Trim your nails regularly. Clip them straight across. File rough edges.

Rest and Sleep

ACTIVITY

Math LINK

Sleep Graph

Keep track of the hours you sleep each night for a week. Make a bar graph of your findings. How many nights did you get ten to twelve hours of sleep?

People your age need ten to twelve hours of sleep each night. Getting enough rest and sleep helps you stay alert in school. It keeps you from being tired and having accidents or getting ill. Here are ways to get plenty of sleep at night.

- **Get plenty of physical activity.** Try to be physically active every day. Walk, bike, or skate.
- **Relax before going to bed.** Read a book, look at a magazine, listen to soft music, or talk with family members.
- **Do not eat or drink caffeine.** Caffeine will keep you awake.
- **Do not go to bed angry or upset.** Talk about your feelings if you are upset or angry.
- **Keep a regular schedule.** Go to bed at the same time each night. Get up at the same time each morning.

Name three things you can do to get plenty of sleep at night.

Getting enough sleep will help you do well in school.

LIFE SKILLS

CRITICAL THINKING

Be a Health Advocate

Work in a small group. Write a skit about healthy sleep tips. Perform the skit for the class.

1. **Choose a healthful action to communicate.** As a group, choose an action that will help you get the proper rest.
2. **Collect information about the action.** Determine why this action helps people get proper rest. Use your textbook or school library to find information.
3. **Decide how to communicate this information.** Will the skit show a healthful action? Decide what the skit will show.
4. **Communicate your message to others.** Perform the skit in front of the class. Other groups should decide whether the action performed is healthful or not.

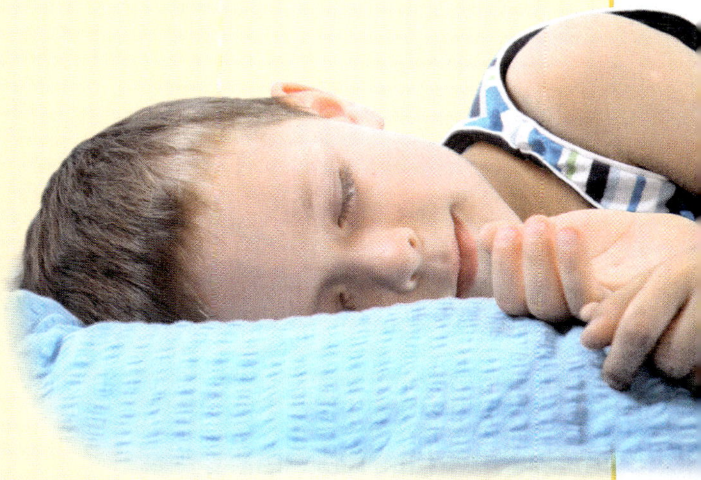

LESSON REVIEW

Review Concepts

1. **Define** the word grooming.
2. **Discuss** how to care for your hair and nails.
3. **Explain** ways to get a good night's sleep.

Critical Thinking

4. **Infer** Monday night you use a new shampoo. Tuesday morning your scalp itches. What could be a reason that your scalp itches?
5. LIFE SKILLS **Be a Health Advocate** Design a poster. Show how to prevent the spread of lice. What information will you include?

LESSON 4

Physical Activity

You will learn . . .

- why being physically active is important.
- how you can work on physical fitness.

Vocabulary

- **physical fitness**, *C21*
- **heart fitness**, *C22*
- **low body fat**, *C22*
- **muscle endurance**, *C22*
- **muscle strength**, *C22*
- **flexibility**, *C22*

Running, biking, and playing tennis are all physical activities. It is important to get plenty of physical activity. When you are physically active, you become physically fit. When you are physically fit, you have energy to work and play.

Benefits of Physical Fitness

Physical fitness means having your body in good shape. Your heart, lungs, bones, joints, and muscles stay strong when you are fit. You can move with ease when you are fit. People become physically fit through regular activity.

Keeping physically fit cuts the risk of disease. It helps you keep a healthful weight. Regular activity improves blood flow to the brain, keeping your mind clear and sharp. You might do better in school, too.

Do You Know?

One type of physical activity is swimming. Swimming is an aerobic exercise. Aerobic exercises help condition the heart and lungs.

How does being physically active improve your mind?

Physical Health Physical activity helps keep your heart, lungs, bones, joints, and muscles strong. It helps you stay at a healthful weight.

Mental and Emotional Health Physical activity can help you manage stress. Being physically fit helps you have a healthful self-concept.

Family and Social Health Making friends that care about physical fitness helps your family and social health. You and your friends can be active together. Your family can be active together.

Kinds of Health Fitness

The five kinds of fitness are listed below.

- **Heart fitness** is the condition of your heart and blood vessels. A strong heart pushes more blood out each time it beats. Between beats, a strong heart has longer to rest. Biking and running improve heart fitness.

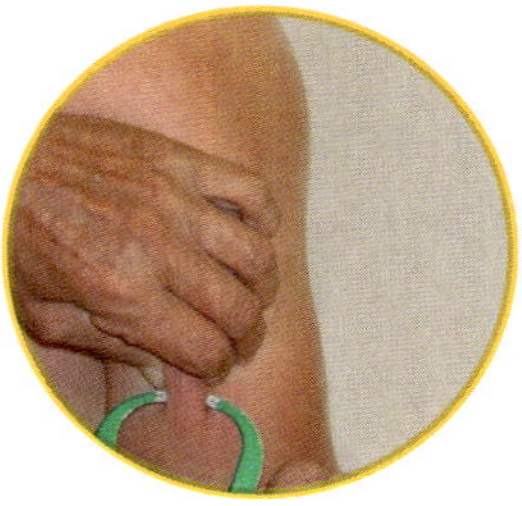

- **Low body fat** is having a lean body without too much fat. Body fat is tested with skinfold calipers (KA•luh•puhrs). These are used to squeeze fat on your arm. The thickness tells how much body fat you have. Swimming and running help lower body fat.

- **Muscle endurance** is the ability to use your muscles for a long time. Swimming laps, walking, and running can improve muscle endurance.

- **Muscle strength** is the ability of your muscles to lift, pull, kick, and throw. Sit-ups and pull-ups help make your muscles strong.

- **Flexibility** is the ability to bend and move easily. Stretching your muscles helps them become more flexible. Gymnastics and dancing are two activities that require flexibility.

What If You Get Injured?

If you get injured in an activity, tell an adult right away. They can help you treat the injury. You may get a cramp in your muscle. A *cramp* is when your muscle contracts or becomes tight. You can get a cramp if your body loses too much water or if you overuse a muscle. If you are tired or feel pain during an activity, take a break and rest.

What should you do if you feel pain during a physical activity?

ACTIVITY LIFE SKILLS CRITICAL THINKING

Manage Stress

Today Latisha is running a race. She is feeling stress. Work with a group. Have one person write down ideas as the group talks about the steps below.

1. **Identify the signs of stress.** What are some signs of stress that Latisha might be experiencing?
2. **Identify the cause of stress.** What might be causing her stress other than just running the race?
3. **Do something about the cause of stress.** What are some things Latisha can do to prepare for her race?
4. **Take action to reduce the harmful effects of stress.** What are some healthful behaviors Latisha can do in the future? Have your group share your ideas with the class.

LESSON REVIEW

Review Concepts

1. **List** the five types of fitness.
2. **Identify** a physical activity that promotes each of the five types of fitness.
3. **Discuss** two benefits of doing physical activity.

Critical Thinking

4. **Explain** the health benefits of being physically fit.
5. LIFE SKILLS **Manage Stress** You have studied hard for tomorrow's math test. But you are feeling stress about taking it. What can you do to manage stress?

LESSON 5

A Good Workout

You will learn . . .

- how to improve fitness skills.
- what steps to include in a physical fitness plan.

Working out means doing physical activity to improve your physical fitness. An example of a workout might include walking for 10 minutes, riding a bike for 30 minutes, and then stretching for 10 minutes.

Vocabulary

- **physical fitness plan**, *C27*
- **warm-up**, *C28*
- **cool-down**, *C28*

Working on Fitness Skills

Fitness skills help do physical activities. They help you workout.

- **Agility** (uh•JIL•i•tee) *Agility* lets you change directions quickly. Playing basketball, soccer, or tennis helps with agility.
- **Balance** *Balance* keeps you from falling. Practice balance by riding your bike or doing martial arts.
- **Coordination** (koh•awr•duh•NAY•shun) *Coordination* lets you use more than one body part at a time. You use coordination when you kick a ball.
- **Reaction Time** The time it takes your muscles to respond is *reaction time*. Picture a batter hitting a ball. How quickly can you catch the ball to get the batter out?
- **Speed** *Speed* is the distance you move in a certain amount of time. As a fitness skill, *speed* means the ability to move fast. When you run you are using this fitness skill.
- **Power** *Power* is the ability to use strong muscles. When you throw a ball fast and far, you are using the power of your arm muscles.

What fitness skill do you use to keep from falling?

Consumer Wise — ACTIVITY

Games That Improve Fitness Skills

Playing games is a good way to build fitness skills. You don't need expensive equipment for all games. For example, all you need to play hopscotch is a piece of chalk, a beanbag, and flat pavement in a safe area. Make a list of games you can play with things you already have?

▲ **Jumping rope builds agility, balance, and coordination.**

Workout

ACTIVITY

Health Online

Men, Women, and Body Fat

A physically fit person has low body fat. Research and report on the relationship between gender and body fat using the e-Journal writing tool.

What is your favorite way to workout? Do you like to run, bike, or swim? Maybe you like to dance or do gymnastics. You can keep track of your workouts in your physical fitness plan. Choose physical activities that you enjoy for your physical fitness plan. You will be more likely to stick to your plan.

Remember to choose physical activities that include the five types of health fitness. The chart below lists each type of health fitness. It also shows examples of activities for each type of fitness.

▼ **People choose activities based on skills, age, and things they like.**

Fitness Ratings for Different Activities

physical activity	heart fitness	low body fat	muscle endurance	muscle strength	flexibility
swimming	♥	♥	♥		
running	♥	♥			
bicycling	♥	♥			
sit-ups			♥	♥	
gymnastics					♥

Your Physical Fitness Plan

You can make a physical fitness plan to keep track of your physical activity. A **physical fitness plan** is a written plan of physical activities. You can use the plan to help you stay fit. You can keep track of the activities you do each week.

Steps to make your physical fitness plan:

1. Make a calendar for one week.
2. Make time on three to five days to work on heart fitness and low body fat.
3. Make time on two to four days to work on muscle strength and endurance.
4. Make time on two to three days to work on flexibility.

How many days should you work on muscle strength and endurance?

▲ **Drink lots of water and sports drinks when it is hot outside so that you don't get dehydrated.**

▼ **Here is a sample physical fitness plan.**

Physical Fitness Plan

Sunday	Monday	Tuesday	Wednesday	Thursday	Friday	Saturday
Muscle strength: Do 15 pull-ups	Heart Fitness: Ride bike for 30 minutes	Flexibility: Practice gymnastics for 20 minutes	Heart Fitness: Jog for 30 minutes	Muscle strength: Do 30 sit-ups	Heart Fitness: Ride bike for 30 minutes	Flexibility: Dance for 30 minutes
Muscle Endurance: Walk for 30 minutes	Low Body Fat: Jog for 20 minutes		Low Body Fat: Ride bike for 20 minutes	Muscle Endurance: Swim 10 laps	Low Body Fat: Jog for 20 minutes	

Warm-Up and Cool-Down

Every workout should include a warm-up and a cool-down. A **warm-up** is three to five minutes of easy physical activity done before a workout. This could include knee lifts, arm circles, walking, or slow jogging. A warm-up helps prevent muscle injuries. It prepares your muscles for physical activity. A **cool-down** is five to ten minutes of easy physical activity done after a workout. Walking and stretching might be part of a cool-down. A cool-down helps your muscles relax after a workout.

Why is it important to warm-up before you begin physical activity?

ACTIVITY

On Your Own

FOR SCHOOL OR HOME

Choosing Physical Activities

There are many activities you can do with your family. Some families might enjoy hiking in a park or flying kites. What physical activities do you enjoy doing with your family? Make a calendar that includes physical activities your family can enjoy together.

After you run, you should do cool-down exercises to relax your muscles.

ACTIVITY

LIFE SKILLS

CRITICAL THINKING

Practice Healthful Behaviors

Work with a partner. Write a television commercial about jogging. State that jogging is a healthful behavior.

1. **Learn about a healthful behavior.** Use your textbook or school library to learn about jogging. Find out how jogging improves a person's health.
2. **Practice the behavior.** In the commercial tell how jogging helps to improve health. Suggest places to jog.
3. **Ask for help if you need it.** Show your commercial to your physical education teacher. Have him or her give you advice on how to make your commercial better.
4. **Make the behavior a habit.** Present your television commercial to the class. Discuss how each student can make jogging a habit.

LESSON REVIEW

Review Concepts

1. **Describe** a physical fitness plan.
2. **List** three physical activities that you can do at home with your family to stay in shape.
3. **Identify** physical activities that you can you do to improve agility.

Critical Thinking

4. **Analyze** how warming up and cooling down prevent injury.
5. LIFE SKILLS **Practice Healthful Behaviors** Name a healthful behavior you can practice before a workout to prevent injury.

LESSON 6

Safety and Physical Activity

You will learn . . .

- ways to prevent injuries during physical activity.
- what safety equipment is needed for different sports.
- ways to be a good sport.

Vocabulary

- **safety equipment**, *C32*
- **mouthguard**, *C32*

Games and sports are more fun if you stay safe while playing them! That's why safety equipment is needed for some sports. There are other ways to prevent you and others from being injured, too.

Ways to Prevent Injuries

There are ways to prevent injuries during physical activities.

Follow the Rules Rules help keep you and others from injury. Respect other players in every activity. Do not hit or push other players. Following rules helps you build healthful relationships.

When to Stop Stop playing if you are tired or have an injury. Tell your coach, parents or guardian. If you keep playing, you could make an injury worse.

Choose a Safe Place to Play Do not play where there may be broken glass or holes. Do not play where there may be moving cars or trucks, or people who might try to harm you. Tell your parents or guardian where you will be playing. A coach or another adult should watch you play sports.

ACTIVITY

BUILD Character

Building a Safe Playground

Citizenship You are building a community playground. What will you include in your design? How will you make it safe? What equipment will you include? Write down your ideas. Draw the playground.

▲ Safe equipment helps keep children safe while they play.

Why should you follow the rules of the game or sport you are playing?

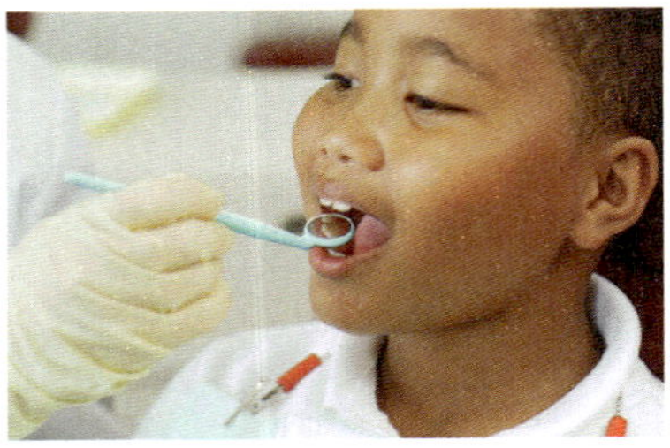

▲ **Up to one-third of all dental injuries happen during sports. Wear a mouthguard during sports in which you might fall or be hit in the mouth by a person, ball, or puck.**

Safety Equipment

Safety equipment protects you from injuries when you play sports or exercise. Do you wear a helmet when you ride your bike? A bike helmet is a type of safety equipment.

Special clothing keeps skin from being scraped, cut, or bruised. Knee pads, elbow pads, and wrist pads protect bones and joints from injury. Safety glasses protect the eyes. In some sports, boys need to wear a protective cup.

A **mouthguard** is an object worn to protect the mouth and teeth. Some sports require all athletes to wear a mouthguard.

▲ **Football** Wear a helmet, pads, and mouthguard to protect your head, body, and mouth.

▲ **Ice Hockey** Wear a helmet, padded guards, gloves and a face mask.

▲ **Soccer** Wear shin guards.

To protect you from injury, safety equipment has to fit. Gear that is too small is uncomfortable. You may not be able to move easily. Gear that is too large may move or come off. Gear should be worn the proper way, with all straps tightened, fastened, and secure.

What can happen if your safety gear does not fit properly?

▲ **Baseball** Wear a helmet when at bat. Catchers wear a helmet, face mask, knee and chest guards, and a protective cup.

▲ **Biking** Wear a helmet and shoes with closed toes. Ride a bike with reflectors on the frame, wheels, and pedals.

▲ **Inline Skating** Wear a helmet, elbow and knee pads, and wrist guards.

Be a Good Sport

You need more than fitness skills to play sports. You need to get along with your teammates. You need to cooperate with them. To *cooperate* is to work well with others toward a goal. This is one way to be a good sport.

There are other ways to be a good sport. Show respect for your teammates. Respect players on other teams, too. Never hit, shove, or kick other players. Do not use wrong language or tease other players. Follow the rules of the game. Listen to the coach or team leader.

Shaking hands with members of the other team is one way to be a good sport.

Be a good sport when you are a fan, too.

If you lose, you may feel disappointed. But don't feel sad or blame others for losing. Work harder to do better next time. Everyone will enjoy the game if you all have fun as you play.

Wait your turn when you are sharing at a playground. Follow rules on the safety signs. Sometimes there are rules for how to use a diving board or bike path. These rules help keep you and others safe.

List two ways that you can be a good sport when you play with others.

ACTIVITY LIFE SKILLS **CRITICAL THINKING**

Resolve Conflicts

Role-play a conflict between Timmy and Jamal. Each role-player will have a partner who will help resolve conflict.

1. **Stay calm.** Timmy pushes Jamal as he goes for a lay up. Jamal's partner tells him how to respond.
2. **Talk about the conflict.** Jamal does what his partner says.
3. **List possible ways to settle the conflict.** Jamal and Timmy's partners' tell them what to say or do.
4. **Agree on a way to settle the conflict. You may need to ask a responsible adult for help.** Jamal and Timmy do what their partners say. The class decides if Jamal and Timmy chose the best way to resolve the conflict.

LESSON REVIEW

Review Concepts

1. **Explain** why it is important to choose a safe place to play sports and games.
2. **List** three examples of safety equipment.
3. **Explain** ways you can be a good sport.

Critical Thinking

4. **Summarize** ways different kinds of safety equipment helps protect you from injury.
5. LIFE SKILLS **Resolve Conflicts** Half the students in your class want to play basketball, while the other half want to play volleyball. Both games cannot be played on one court at the same time. List possible ways to settle the conflict.

Learning **LIFE SKILLS**

CRITICAL THINKING

Use Communication Skills

Problem Sam, Justin, and Desmond want to play a game. They are all talking at the same time.

Solution It is important to be able to communicate well while playing games. The friends decide to use communication skills.

Foldables™ To Learn Life Skills

Learn This Life Skill

These steps will help the friends communicate. The Foldables™ can help.

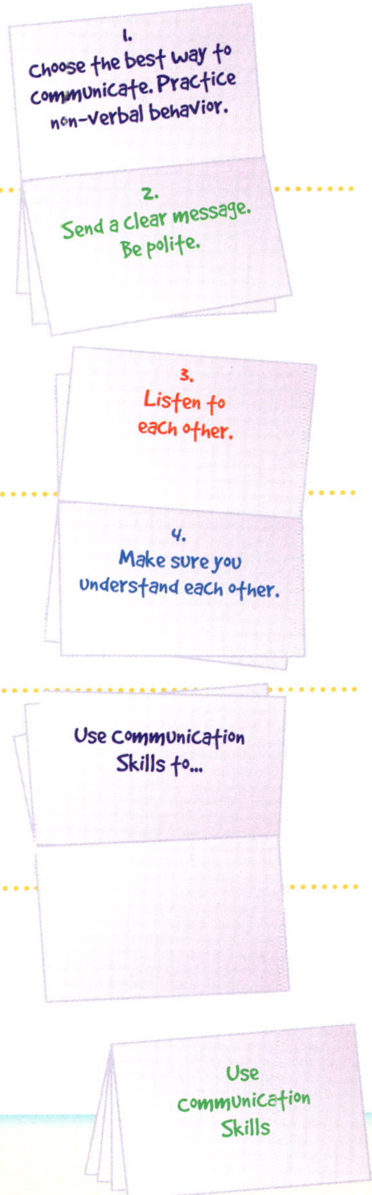

1 Choose the best way to communicate. Practice non-verbal behavior.

How can the friends talk about the type of games they might play?

2 Send a clear message. Be polite.

How can the friends send a clear message?

3 Listen to each other.

The friends listen quietly to each other.

4 Make sure that you understand each other.

The friends review what was said. What else could they do?

Practice This Life Skill

Work in groups of six. Have everyone speak at the same time. What happened? Now try taking turns. How did communication improve?

CHAPTER 5 REVIEW

dental plaque, *C6*
flexibility, *C22*
grooming, *C16*
health-care product, *C16*
mouthguard, *C32*
muscle endurance, *C22*
physical fitness, *C21*
warm–up, *C28*

Use Vocabulary

Choose the correct term from the list to complete each sentence.

1. You should always __?__ at the beginning of a workout.
2. Taking care of your body and appearance is called __?__.
3. Having your body in top condition is called __?__.
4. You can protect your mouth and teeth from injury during physical activity by wearing a __?__.
5. A product used for grooming is called a(n) __?__.
6. The ability to use your muscles for a long time is called __?__.
7. The ability to bend and move easily is called __?__.
8. A sticky material that forms on teeth is called __?__.

Review Concepts

Answer each question in complete sentences.

9. What are five healthful grooming habits you should practice?
10. What does a doctor do during a medical checkup?
11. What should you do if your eyes itch or burn?
12. Why is it important to choose physical activities you enjoy doing for your physical fitness plan?
13. What are three sports that can help you improve agility?

Reading Comprehension

Answer each question in complete sentences.

You rely on your sense of sight, or vision, in many ways. Your eyes help you stay safe. For example, you look up and down the street to check for cars before crossing. Your eyes also help you communicate. You show respect to someone if you look at them while you are talking.

14. What does *vision* mean?
15. What is one way that your eyes help you stay safe?
16. How do your eyes help you communicate with others?

Critical Thinking/Problem Solving

Answer each question in complete sentences.

Analyze Concepts

17. After you get to the hockey rink, you realize you forgot your safety equipment. Explain how not wearing safety gear might cause injury.

18. At football practice, you find that your helmet is too big. Explain why it is important for your safety gear to fit properly.

19. You are accidentally hit with a pitch. Tell why it is important to be a good sport at such times.

20. Explain how lack of sleep affects your health. Name two ways you can get plenty of sleep each night.

Practice Life Skills

21. **Use Communication Skills** Your older sister is playing loud music. You cannot do your homework. How can you use communication skills to solve this problem?

22. **Make Responsible Decisions** You don't have your safety equipment with you, but your friends want you to play soccer anyway. What should you do?

Read Graphics

The chart lists the decibels of five sounds. Decibels measure how loud or soft a sound is. Use it to answer questions 23–25.

Sound	Decibels
Whisper	30
Talking in a normal voice	60
Lawnmower	90
Chainsaw	100
Jet plane engine	140

23. Sounds louder than 85 decibels can cause hearing loss. Which of the sounds on the chart could cause hearing loss?

24. Calculate how much higher in decibels a chainsaw is than your normal talking voice.

25. Decide if the decibels of a lawnmower are lower or higher than the decibels of a jet plane engine.

CHAPTER 6

Violence and Injury Prevention

What Do You Know?

You can stay safe and have fun at the same time. Look at the list of safety equipment below. Which equipment do you wear when you bike, skate, or ride a scooter?

_?_ helmet
_?_ wrist pads
_?_ elbow pads
_?_ knee pads

Did you check all of the equipment? If so, you have been practicing safety rules. Read **Violence and Injury Prevention** to learn more about safety.

LESSON 1

Indoor Safety

You will learn . . .

- how to prevent falls.
- how to prevent fires.
- ways to avoid injury from poisons.
- what computer safety means.

Vocabulary

- **accident**, *C43*
- **injury**, *C43*
- **hazard**, *C43*
- **fire escape plan**, *C45*
- **poison**, *C47*

Suppose you leave your toys in the middle of the floor. Someone could trip over them and get hurt. Remember to put your toys away to prevent injuries. This is an important safety rule. Following safety rules can help keep you safe.

Ken Karp/McGraw-Hill Education

Follow Safety Rules

An **accident** is something that is not supposed to happen. An **injury** is damage or harm done to a person. Most injuries are caused by accidents.

Suppose someone leaves a skate on the floor. You trip on the skate. The skate is a hazard. A **hazard** is something that can cause harm or injury. The following safety rules can help you prevent accidents.

How to Prevent Accidents

At Home You can take steps to help your parents or guardian make home a safe place.

- **Put away toys, books, and other objects.** These items are tripping hazards.
- **Use a stepstool to reach things up high.** First get permission from an adult.
- **Try not to walk on wet floors.** Wet floors are slippery and dangerous.

At School You can follow safety rules in school, too.

- **Keep the floor free of objects.** That way you won't make a tripping hazard.
- **Walk in the hallways.** Watch for other people so that you don't bump into them.

Why is it important to follow safety rules?

ACTIVITY

Art LINK

Draw a Danger Sign

Cut a large circle out of a sheet of paper. Think of a hazard. Then draw a sign that warns people about the hazard. Color your sign and share it with your family and classmates.

▲ Try not to walk on wet floors.

▲ **STOP! Do not run. That can make things worse.**

▲ **DROP to the ground.**

▲ **ROLL back and forth to put out the flames. Cover your face with your hands.**

Safety Rules to Prevent Fire

You can help keep your home safe from fire.

- **Have your parents check smoke alarm batteries.** Smoke alarm batteries should be checked at least twice a year.
- **Call for help if you see a fire.** Dial 9-1-1.
- **Practice Stop, Drop, and Roll.** If your clothes catch on fire, you should stop, drop, and roll. This will put the fire out.

Electricity: Play it Safe

Use an electric appliance only if your parent or guardian is there to help you.

Electrical cords and extension cords should be used safely. Don't use appliances if their cords are damaged. Don't put them under carpets or across doorways.

Don't plug more appliances into an outlet than it can hold. The outlet could heat up and catch fire.

In Case of Fire

You and your family should have a fire escape plan. A **fire escape plan** is a map of your home that shows different ways out of every room. Your plan should include two ways out of each room and a place to meet your family outside.

What should you do if you are trapped inside a room during a fire?

ACTIVITY

Social Studies LINK

Be a Role Model

Contribute to your school community. Teach younger students about fire safety. Make fire-safety materials to show them. The materials could be fire-safety coloring pages, fire-safety comic strips, or fact sheets for kids to take home. The information you give could save a younger student's life.

▲ The first way out of a fire should be a door. Feel a door before you open it. If it is hot, don't open it. If the door is cool, open it a crack. Check for smoke.

▼ If it is smoky, crawl on your hands and knees to keep below the line of smoke. Cover your mouth and nose. If you can't do these things, close the door and cover the bottom with a towel. Go to a window and yell for help.

What Is Computer Safety?

ACTIVITY

On Your Own

FOR SCHOOL OR HOME

Surfer Safety Reminder

RULE: For safety's sake, never surf the Internet alone. To remember this rule, make a surfer T-shirt. Ask your parents or guardian to buy you a plain white T-shirt. With a permanent marker, write across the front of the T-shirt: *Surf With A Parent Only.* Decorate your T-shirt with permanent markers of different colors.

People use computers to write reports and do research on the Internet. The Internet has sites that are not safe for children. Also, strangers may be able to find out private information about you on the Internet. Play it safe when using the Internet. Here are some tips for you.

- **Use the Internet** only when your parent or guardian gives you permission.
- **Do not give out** your name, address, telephone number, age, photo, school, parents' names, or other personal information.
- **Tell your parent or guardian** if anyone tries to find out where you live or tries to meet you.

Why is it important not to give out personal information on the Internet?

Only use the Internet when your parent or guardian is with you.

Onoky Photography/SuperStock

Poison Control

A **poison** is a drug or substance that harms or kills. Eating or drinking poison, or breathing in its fumes, is harmful. A poison can also cause harm by entering a person's body through the skin.

A parent or guardian should always help you use household products. Move away and get fresh air if you smell a strong odor. Keep household products in their own labeled containers and away from young children.

How can someone be harmed by a poison?

ACTIVITY

LIFE SKILLS

CRITICAL THINKING

Practice Healthful Behaviors

1. **Learn about a healthful behavior.** Learn how to make a fire escape plan for your home.
2. **Practice the behavior.** Practice your plan with your family. Use the door as the exit when you practice.
3. **Ask for help if you need it.** Have your parents or guardian review your plan. Agree on a place to meet outside.
4. **Make the behavior a habit.** Have home fire drills every month.

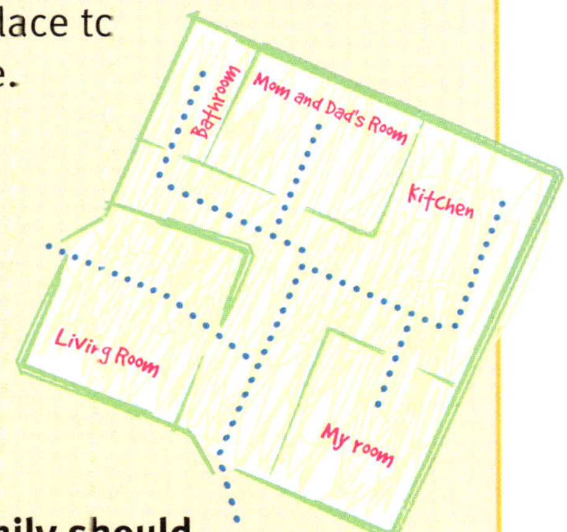

▶ **Every family should have a fire escape plan.**

LESSON REVIEW

Review Concepts

1. **Describe** three ways you can prevent falls at home.
2. **Explain** how you can prevent injuries caused by electricity.
3. **Discuss** why you should use the Internet safely.
4. **Name** two ways to protect yourself around poisons.

Critical Thinking

5. **Comprehend** A young child visits your home. The cleaning materials are kept in a low cabinet. How could the child be harmed? How could you help prevent it?
6. LIFE SKILLS **Practice Healthful Behaviors** You see that your family does not have a fire escape plan. What healthful behavior can you work on to keep safe in case of a fire?

LESSON 2

Safety on the Go

You will learn . . .

- how to keep safe around cars and buses.
- safety rules when walking, biking, skating, and riding a scooter.
- how to keep safe around water.

Vocabulary

- **seat belt**, *C49*
- **personal flotation device**, *C52*

A common way to travel is by car. Another way is by bus. Do you ride a bus to school? Do your parents or guardian drop you off in a car? There are safety rules for each of these ways to travel.

Safety on the Go

People use buses and cars to travel from one place to another. Some people take a bus to work. Others use a car. You might ride on a bus to go on a school field trip. Maybe you ride in a car with your family to the grocery store. You should follow safety rules when you ride in buses and cars.

Always wear a seat belt when you ride in a car. A **seat belt** is the lap and shoulder belt worn in a car. Ride in the back seat. It is safer than riding in the front seat. Lock the door when you are seated. Get out on the curb side of the car.

Follow these safety rules when taking a bus.

- **When waiting for a bus** try to take at least four giant steps away from the curb.
- **Never walk behind a bus.** The bus driver can't see you.
- **Cross only in front of the bus.** Wait for the driver to signal that it is okay to cross the street.
- **On the bus, sit quietly in your seat.** Keep aisles clear.

What can you do to keep safe while riding on a bus?

▲ **Find the Right Seat Belt.** Each seat in a car has its own belt. Make sure that you have the right belt for your seat.

▲ **Pull the Belt Across Your Lap.** Sit up straight. Pull the belt from your shoulder down to the opposite hip.

▲ **Snap the Belt Tight Around Your Hips.** Snap the two sections of the belt together. Make sure that the belt is tight around your hips.

Safety and Sports Vehicles

Do You Know

Bike Safety

If everyone wore a helmet when biking, one death each day could be prevented. One brain injury every four minutes could be prevented.

When you walk, skate, bike, or ride a scooter, you should follow safety rules to stay safe.

Skating and Riding Scooters

Wear a helmet, elbow pads, wrist guards, and knee pads when you skate or ride a scooter. Do not skate or ride in traffic. Skate or ride on smooth areas. Do not skate or ride at night. Use hand signals so that other people will know what you are going to do.

Right Turn
Stretch out your left arm, bend at the elbow, and point your forearm up.

Left Turn
Stretch out your left arm, with your forearm extended out.

Stop
Use your left arm. Bend your elbow down. Your palm faces the back to show stop signal.

Riding a Bike

You might ride a bike to school. Or, you might ride your bike with your friends for fun. Follow these rules to stay safe.

- **Always wear a helmet flat on the top of your head.** Make sure that the helmet fits you.
- **Ride your bike on a bicycle path when possible.** If you ride on a street or road, ride on the right hand side.
- **Ride only in daylight.** People cannot see you at night.
- **Stop and look both ways when you cross a street with your bike.** Walk your bike across the street.
- **Don't wear long pants that are loose at the ankle.** They could get caught in the bike and cause you to crash.

ACTIVITY

On Your Own

FOR SCHOOL OR HOME

Make a Poster

Make a poster showing safety rules for riding a bike. Ask for permission to display your poster at home.

Describe a safe area to skate.

▼ **Always follow safety rules when you ride bicycles and skate.**

Water Sports

Playing in or near water can be fun. Follow these safety rules when you swim and boat.

Swimming

Follow these safety rules when you swim.

- **Learn to swim.** Take swimming lessons. A swimming teacher can show you how to swim.
- **Wear sunscreen when you swim outside.** Sunscreen will protect your skin from the Sun.
- **Swim with a lifeguard on duty.** Do not swim by yourself. Swim with a parent, guardian or trusted adult.
- **Swim only where signs say that swimming is allowed.** It is not safe to swim in every area.
- **Do not go in deep water.** Do not go in water that is above your head.

▲ Always wear a personal flotation device when you are on a boat. What other safe practices do you see here?

Riding in a Boat

Follow these rules when you ride in a boat.

- **Always wear a personal flotation device** (PFD), an object that helps you stay afloat in water.
- **Do not hang over the side of the boat.** You could fall out.
- **Stay seated.** It is easy to tip a small boat if you stand up in it.

How can you swim safely?

LIFE SKILLS

CRITICAL THINKING

Make Responsible Decisions

Aza wants to swim in a lake with no one watching. Help Aza make a responsible decision.

1. **Identify your choices. Check them out with your parent or trusted adult.** Write three choices Aza could make on a three-column chart.
2. **Evaluate each choice. Use the *Guidelines for Making Responsible Decisions™*.** Write your responses on the chart.
3. **Identify the responsible decision. Check this out with your parent or trusted adult.** Write the decision at the bottom of the chart.
4. **Evaluate your decision.** Discuss possible outcomes of the decision.

Guidelines for Making Responsible Decisions™

- **Is it healthful?**
- **Is it safe?**
- **Does it follow rules and laws?**
- **Does it show respect for myself and others?**
- **Does it follow my family's guidelines?**
- **Does it show good character?**

LESSON REVIEW

Review Concepts

1. **List** three safety rules for riding in a car.
2. **Explain** how to cross the street safely after leaving a school bus.
3. **Compare** the safety rules for riding a bike and for skating.
4. **Name** three safety rules to follow while riding in a boat.

Critical Thinking

5. **Analyze** A friend wants to swim far out from the beach to see how deep the water is. Is your friend following safety rules? Explain.
6. LIFE SKILLS **Make Responsible Decisions** You see a friend across the street. You want to cross. You are halfway down the street. What is the responsible decision? How could you use the decision making steps to decide?

Safe in Wind and Weather

You will learn . . .

- how to stay safe outdoors.
- how to stay safe in bad weather.

Vocabulary

- **thunderstorm**, *C56*
- **flood**, *C56*
- **earthquake**, *C56*
- **tornado**, *C57*

What activities do you enjoy outdoors? Do you visit the playground? Perhaps you enjoy swinging on the monkey bars. What should you do if the weather suddenly turns bad? You should go inside right away.

Outdoor Safety

When you run and play outdoors, it is important to be safe. Make sure to follow these safety rules. These rules will help keep you safe on or near playgrounds.

- Don't play on equipment that is broken.
- Don't tie ropes to equipment. You could get tangled.
- Don't try stunts or dares.
- Make sure that a responsible adult is nearby.

More Outdoor Safety Rules

- Check that your shoelaces are tied when you run.
- Don't climb trees, telephone poles or power poles.
- Don't run near a swimming pool or on wet ground. You could slip and fall.

Why should you not tie ropes to playground equipment?

ACTIVITY

Physical Education LINK

Hurling Safety Rules

Play with the whole class or with a partner. You will need a beach ball. Stand in a large circle (or facing a partner). Begin by tossing the ball to a player. As the player catches the ball, he or she names an outdoor safety rule. Then he or she tosses the ball to another player. Keep going until all the rules have been named.

▶ Keep your shoes tied so that you don't trip.

Natural Disasters

Natural disasters are events caused by nature. They result in heavy damage. Storms, floods, and earthquakes are natural disasters. You can stay safe in them.

Thunderstorm

A **thunderstorm** is a storm with thunder and lightning. Follow these safety rules to stay safe.

- **Go inside.** You will be safer from lightning when indoors.
- **Do not use the telephone.** Lightning can travel through wires.
- **Keep away from water.** Lightning can travel through some plumbing pipes.

Flood

A **flood** is the overflow of water onto dry land. Do not go into flood water. Listen to the radio for emergency information.

Earthquake

An **earthquake** is a shaking or trembling of the ground. To protect yourself, drop to the floor and get under something for cover, such as a desk, or stand in a doorway. Stay away from glass, windows, and bookcases. Move away from buildings and electrical wires if you are outdoors.

Do You Know?

Thunderstorm Fact

Lightning strikes before rain falls. Safety Tip: Stop activities and take shelter when you see lightning. Do not wait for rain.

▲ **Satellites are used to track harmful weather conditions. This kind of technolgy can help save lives.**

imac/Alamy

Follow Safety Rules for a Tornado

A **tornado** is a powerful storm with winds that whirl in a dark cloud. Follow these safety rules to keep yourself safe during a tornado.

- **Go indoors.** If you can't get indoors, lie in a low area.
- **Go to a basement** or go to the center of a ground floor.
- **Stay away from windows.** The glass could blow out of the window and cut you.

How can you stay safe during a tornado?

ACTIVITY
LIFE SKILLS
CRITICAL THINKING

Access Health Facts, Products, and Services

1. **Identify when you might need health facts, products, and services.** Suppose a bad storm is coming toward your town. You need to know how to protect yourself.
2. **Identify where you might find health facts, products, and services.** Find information about storms using your textbook.
3. **Find the health facts, products, and services you need.** Identify two sources of information about storms.
4. **Evaluate the health facts, products, and services.** Decide which information is the most important to help keep yourself safe from harm.

LESSON REVIEW

Review Concepts

1. **List** three safety rules to follow when you run outdoors.
2. **Explain** how you can stay safe on the playground.
3. **Tell** why you should not use a telephone during a thunderstorm.
4. **Name** three rules to follow during a tornado.

Critical Thinking

5. **Plan** Suppose that you are at a playground and you see that a thunderstorm is coming. Describe the steps you would take to stay safe.
6. LIFE SKILLS **Access Health Facts, Products, and Services** Suppose that your community has very cold weather. How can you find out how to keep safe during this weather?

Learning LIFE SKILLS

CRITICAL THINKING

Set Health Goals

Problem Alex knows that he can get injured at the playground. He wants to set a goal to stay safe while playing.

Health Behavior Contract

Name ______________________ **Date** __________

Health Goal: I will follow safety rules at school and home.

Effect on My Health: I will stay safe from harm and injury if I follow safety rules.

My Plan: I will make a chart that lists the days of the week. I will track each safety rule for playground equipment that I follow. I will use the chart below to keep track of my progress. I will get help from my parents and my teacher.

Safety Rules	S	M	T	W	Th	F	S
I will make sure an adult is nearby.	×			×	×		×
I will not play on equipment that is broken.		×		×		×	×
I will only play on equipment with a soft surface underneath.			×		×	×	×

How My Plan Worked: My plan helped me to stay safe while playing on the playground. I followed all of the safety rules. I kept safe from injury and harm.

Solution Alex can figure out what goal he wants to set about playground safety. Then, he can make a plan to follow using the steps on page C59.

Foldables™ To Learn Life Skills

Learn This Life Skill

Follow these steps to help you set health goals. The Foldables™ can help you.

1 Write the health goal you want to set.

Alex set a health goal: I will follow safety rules for home and school.

2 Explain how your goal might affect your health.

Alex knows that following playground safety rules can prevent injuries. What injuries can Alex prevent?

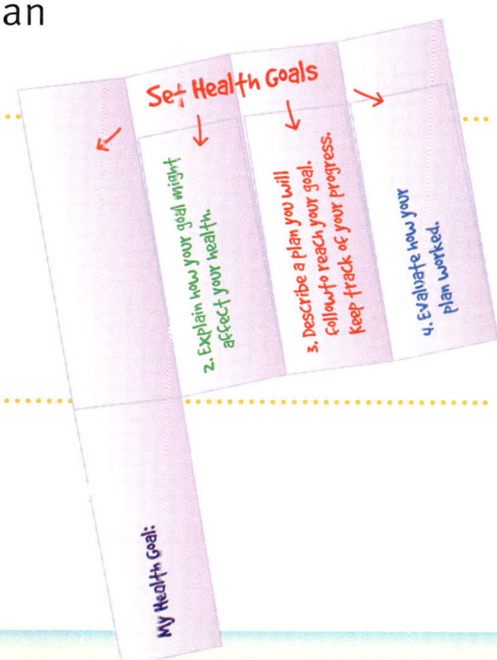

3 Describe a plan you will follow to reach your goal. Keep track of your progress.

Alex makes a chart to keep track of his progress. How did Alex do?

4 Evaluate how your plan worked.

Alex tried to improve each day. What could Alex do to make his plan better?

Practice This Life Skill

Role-play. Set a goal for following safety rules when you play. Make a health behavior contract.

LESSON 4

Staying Safe Around People

You will learn . . .

- how to stay safe at home.
- how to stay safe from strangers.
- what is an unsafe touch.

Vocabulary

- **violence**, *C61*
- **stranger**, *C62*
- **unsafe touch**, *C63*

You have friends you play with at school. Teachers and other trusted adults are there to help you. But what should you do when you meet strangers? You can follow rules to stay safe.

Keep Safe at Home

Sometimes your parents or guardian can't answer the phone or door. They might be busy cleaning or helping another family member. Or, someone might call your home and ask questions. Sometimes these people use violence to get what they want. **Violence** is harm done to yourself, others, or property. To keep yourself safe, follow these safety rules if someone comes to your home.

- **Keep the doors locked.** Stay inside.
- **Don't let anyone into your home.** Wait until your parents or guardian can come to the door. If you can't find your parents or guardian and the person won't go away, dial 9-1-1.
- **Don't give your name, address, or telephone number to someone on the telephone.** Do not answer any questions. Hang up the phone if the caller bothers you.

What do you do if someone you do not know calls your home?

Write About It!

Dialogue Look at the pictures below. Write sample dialogues like those you see in the pictures. Work with a partner to practice the situations and lines that you wrote.

▲ Always tell a caller that your parents or guardian cannot come to the phone if they are busy.

Strangers and Your Safety

ACTIVITY

Health Online

Amber Alert System

The Amber Alert system helps recover missing children across the United States. Research and report on how the Amber Alert system helps find missing children using the e-Journal writing tool.

A **stranger** is someone you do not know well. Keep safe around strangers by following these rules.

- Do not play alone.
- Do not take candy, money, or toys from a stranger.
- Do not go near a stranger who wants to show you a pet.
- Never get into a car with a stranger.
- Do not go anywhere with a stranger, even if a stranger says that your parents or guardian said it's okay.

If a stranger bothers you, run away. Go to where there are other people, such as into a store. Yell so that people will know you are worried and will help. Tell your parents or guardian about a stranger who bothers you.

Who Can You Trust If You Are Lost?

If you are ever lost or don't know where you are, do not panic. Read the following list to see which people you can trust.

- An adult who works in a store
- A lifeguard if you are lost at the beach
- A teacher at your school
- A police officer or firefighter

▲ **You can trust a teacher if you are lost.**

A Touch That Is Not Right

An **unsafe touch** is a touch that is wrong. It is not okay for anyone to touch your private body parts. To protect yourself, you can:

- **Tell** the person to stop touching you.
- **Run away** from this person.
- **Yell** as loud as you can.
- **Tell a responsible adult.** If the first adult you tell does not believe you, tell another adult.

What is an unsafe touch?

ACTIVITY LIFE SKILLS **CRITICAL THINKING**

Use Resistance Skills

Act out a skit. A stranger comes to your door wearing a uniform. He asks to come in.

1. **Look at the person. Say "no" in a firm voice.** The uniformed man knocks on the door. The student says "no" when the uniformed man asks to come in.
2. **Give reasons for saying "no."** The student's parents are busy cleaning.
3. **Match your behavior to your words.** The student makes sure the door is locked and walks away.
4. **Ask an adult for help if you need it.** The uniformed man won't leave. The student uses the telephone to dial 9-1-1.

Students discuss the skit. They decide whether the student handled the uniformed man correctly.

LESSON REVIEW

Review Concepts

1. **Explain** what to do to keep safe if someone comes to the door of your home.
2. **List** three safety rules to stay safe from strangers who might want to harm you.
3. **Explain** how to protect yourself from an unsafe touch.

Critical Thinking

4. **Evaluate** You are waiting for your parent to pick you up after school. A person comes and says that your parent sent him to pick you up. Is it safe to go with the person? Explain.
5. LIFE SKILLS **Use Resistance Skills** Suppose that a stranger in a car asks you for help. How can you use resistance skills to keep safe?

LESSON 5

Staying Safe From Violence

You will learn . . .

- ways that you can stay safe from a gang.
- why you should not pretend to have a weapon.
- what to do if you find a weapon.

Vocabulary

- **gang**, *C65*
- **weapon**, *C66*

Young people should make friends who are safe to be around. Friends should help each other practice healthful behaviors.

Gangs: What You Should Know

What do you know about gangs? A **gang** is a group of people involved in acts that are dangerous or against the law.

Gang members often wear certain colors or clothes. They might draw graffiti. Gang members act in violent ways. They often carry weapons, such as knives and guns. They sometimes use weapons to hurt people. Gang members often bully others to join or to obey. They might sell and use drugs, and rob other people. Gang members often fight members of other gangs.

BUILD Character ACTIVITY

A Real Family

Caring Draw a cartoon about Joey. Joey is a young man who is in a gang. His friends convince him to leave the gang. Joey becomes closer to his real friends.

Why do some people join gangs?

▶ True friends care for one another.

Weapons: Know What to Do

Write About It!

Write a Song Write a song about the importance of following safety rules and staying away from weapons. Read or sing your song for the class.

A **weapon** is an object that is used to harm someone. A weapon can be a gun, knife, or a baseball bat. A person might be harmed on purpose by someone with a weapon. Sometimes a person is around a weapon and is harmed by accident.

It is important to protect yourself from weapons. Follow these safety rules.

- **Do not touch or carry a weapon.** You could be accidentally injured.
- **Do not pretend to have a weapon.** Someone could harm you if they think you are threatening them.
- **Do not bring weapons or toy weapons to school.** It is illegal.

Weapons at Home

You might see a weapon in your home or a friend's home. Never touch the weapon. Talk to your parents or guardian about safe storage of weapons. A gun can be a weapon. It should be locked in a gun case. Guns should not be loaded. They should have triggers locked. Ammunition should be stored by itself.

How should a gun be stored?

S.A.F.E.

If you see or find a weapon, think **S.A.F.E.**

S—**S**top.

A—**A**void going near the weapon.

F—**F**ind an adult.

E—**E**xplain what you saw.

S.A.F.E.

ACTIVITY

LIFE SKILLS

CRITICAL THINKING

Analyze What Influences Your Health

Role-play a conversation between Arlene and Sammy about a knife in a drawer.

1. **Identify people and things that can influence your health.** Arlene shows the knife to Sammy. The class decides who is trying to influence Sammy.
2. **Evaluate how these people and things can affect your health.** Arlene asks Sammy if he wants to hold the knife.
3. **Choose healthful influences.** Sammy explains his decision to Arlene.
4. **Protect yourself against harmful influences.** The class discusses how Sammy can protect himself from harmful influences.

LESSON REVIEW

Review Concepts

1. **Analyze** how members of gangs act.
2. **Tell** why you should not pretend to have a weapon.
3. **Name** three rules that protect you from weapons.
4. **Explain** what the letters **S.A.F.E.** mean if you find a weapon.

Critical Thinking

5. **Decide** At school, your friend tells you that he has a weapon in his book bag. Decide what you should do. Tell why your friend is acting unsafe.
6. LIFE SKILLS **Analyze What Influences Your Health** Suppose that your older brother shows you a pocket knife. He asks you if you want to play with it. Evaluate how your brother is influencing your health.

LESSON 6

Emergencies

You will learn . . .

- how to call for help in an emergency.
- how to help an injured person.
- what safety rules to follow in case of a disaster.

Vocabulary

- **emergency**, *C69*

Emergencies can happen anywhere at anytime. You could be at home during bad weather. You might be at school when someone trips and cuts his or her knee. You might be playing with your friends outside when someone gets injured. It is important to know what to do in case of an emergency.

U.S. Air Force photo by Maj. Jon Quinlan

What Is a Medical Emergency?

An **emergency** is a situation in which help is needed quickly. It could be caused by bad weather, a fire, or crime. A medical emergency is when someone's health has been harmed. The following situations are medical emergencies:

- breathing is hard or stops
- having a bad chest pain
- swallowing a poison
- bleeding that will not stop
- coughing or spitting up blood
- feeling very bad pain
- passing out

Use the steps listed at the right when you need to call for help.

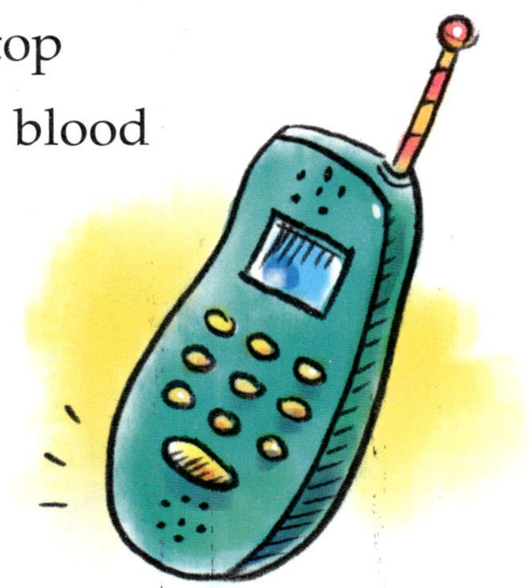

Helping an Injured Person

1. **Stay calm.** Try to think clearly.
2. **Stay safe.** Make sure the area around you is safe so that you do not injure yourself.
3. **Never** move an injured person.
4. **Get help fast!** Get an adult or tell someone to do so. Dial 9-1-1 if it is an emergency.
5. **Keep the person warm.** Use a blanket or jacket.

What are the rules for helping a person who is injured?

Being Prepared

Do you know what to do during an emergency? You can make a disaster kit. Include first aid supplies, a flashlight, and a battery powered radio. You should also include food, water, and a plan for how to get in touch with family members in an emergency. Keep a list of phone numbers. If you need to call for help, dial 9-1-1. Tell what the emergency is. Give your name and the address where you are. Listen to what you are told to do. Stay on the phone until you are told to hang up.

CAREERS

Emergency Medical Technician

An emergency medical technician (EMT) gives emergency medical care to people who are injured or suddenly ill. An EMT might respond to emergency calls, such as accidents. He or she might treat wounds, bleeding, broken bones, breathing problems, and heart attacks.

▲ A family disaster kit is very important to have in case of a disaster.

Safety Drills

All schools have drills that they practice in case of disasters. Follow these safety rules during drills to protect yourself.

- **Know your alarm sounds** to recognize what the different alarm sounds mean.
- **Stay quiet.** Listen to your teacher's instructions.
- **Go to the place that your teacher and school have decided on.** Be orderly. Do not run.

How can your family be prepared for a disaster?

ACTIVITY LIFE SKILLS CRITICAL THINKING

Manage Stress

You are thinking about how to be safe in an emergency. You are feeling stress. What can you do? Work with a partner. Answer the questions below. Write your answers on the chalkboard.

1. **Identify the signs of stress.** What are the signs of stress you might have?
2. **Identify the cause of stress.** What types of emergencies might cause you stress?
3. **Do something about the cause of stress.** What can you do to relieve your stress?
4. **Take action to reduce the harmful effects of stress.** What action can you take to reduce the stress you feel about emergency situations?

LESSON REVIEW

Review Concepts

1. **Describe** three examples of times when you need to call for emergency medical help.
2. **Explain** what you should do to help yourself think clearly while helping an injured person.
3. **List** three items to include in a disaster kit.

Critical Thinking

4. **Analyze** Suppose that you want to have drills for natural disasters. Which natural disasters are likely to happen where you live? How do you know?
5. LIFE SKILLS **Manage Stress** You are having stress about what to do for your pet in case of a disaster. What can you do to manage your stress?

LESSON 7

How to Give First Aid

You will learn . . .

- what precautions to take when giving first aid.
- how to give first aid for cuts, scrapes, nosebleeds, bruises, burns, insect stings, animal bites, reactions to poisonous plants, and choking.
- what items should be part of a first aid kit.

Vocabulary

- **first aid**, *C73*
- **first aid kit**, *C77*

A person who is injured might need your help. He might have a cut on his elbow. Maybe she has an insect bite on her arm. You can use first aid to help improve a person's health.

Comstock Images/Jupiter Images/Alamy

What Is First Aid?

Sometimes people have accidents. They might fall on the cement and scrape their knees. They might burn their finger on the stove. Maybe they stub their toe on a chair. These injuries are minor injuries. They can be treated by giving first aid. **First aid** is the quick care given to a person who is injured or is suddenly ill. You can use first aid steps to treat many minor injuries. There are some steps you should always follow when you perform first aid.

Follow these steps when performing first aid to protect yourself.

1. **Do not touch someone else's blood.** You could get someone else's germs or give your germs to someone else.
2. **Wash your hands with soap and water after giving first aid.**
3. **Do not eat or drink while giving first aid.**
4. **Do not touch your mouth, eyes, or nose while giving first aid.**
5. **Ask an adult if you need help.**

Why should you avoid touching someone else's blood?

ACTIVITY

On Your Own

FOR SCHOOL OR HOME

Emergency Telephone Numbers

If someone gets injured and needs emergency care, every second counts. Having a list of telephone numbers can save time. Make a list of emergency and important telephone numbers to post by the phone. Include your home address so that it is handy.

EMERGENCY TELEPHONE NUMBERS
3747 Maple Drive

Mom at work: 555-3829
Dad at work: 555-8593
Grandma: 555-7075
Aunt Kathy: 222-0536
Police Station: 555-1954
Fire Station: 555-7462
OTHER: 9-1-1

Steps in Giving First Aid

Minor cuts, scrapes, burns, and bruises are minor injuries. Many injuries can be treated with first aid. Follow these steps when treating injuries.

Animal Bite

- **Tell an adult if you get a bite.** Some ticks and mosquitoes carry diseases.
- **Clean the area with soap and water.** Put ice over it.
- **If you get bitten by a dog, cat, raccoon, skunk or a snake, get emergency care right away.** Some animals carry a harmful disease called rabies.

Nosebleed

- **Tell an adult right away.**
- **Lean your head forward.**
- **Pinch your nostrils shut** for ten minutes. Do not blow your nose.
- **Breathe through your mouth.** Spit out any blood.
- **Sit upright.**

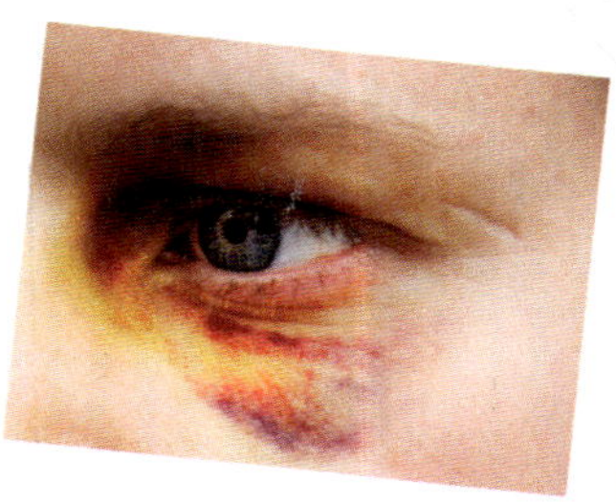

Bruising

- **Have a parent or guardian take you to a doctor** if the bruise does not go away within two weeks.
- **Place a cold cloth** over the area.

Sting

- **Tell an adult if you get a sting.** If the insect has a disease or poison, you will need emergency care.
- **If you get a sting, have an adult remove the stinger.** Scrape it out with a hard card or a clean nail file. Do not squeeze it out.
- **Clean the area with soap and water.** Put ice over it.
- **If you can't breathe, if you feel dizzy, or if the area swells,** get emergency care right away.

Cuts and Scrapes

- **Press on the cut** with a bandage if the cut bleeds.
- **Clean the cut** with soap and water.
- **Cover the cut** with a clean bandage.
- **If the bleeding does not stop,** get a responsible adult or dial 9-1-1.

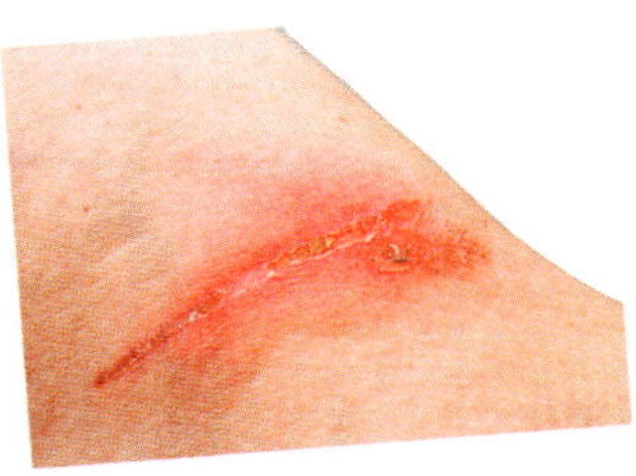

Burn

- **Put a cold cloth over the burn,** or run cold water over the burn for 10 minutes.
- **Cover the burn** with a clean bandage.
- **Do not break** a blister or bubble.
- **If the burn or a blister gets infected,** see a doctor. There are signs to look for if a blister is infected. An infected blister may become warm or swell up. It may become redder or the fluid may drain. It may become more painful.

Rash

- **If the rash is in or near your eyes,** see a doctor.
- **Use calamine lotion or cool water** to treat the rash.
- **Do not scratch or rub** the rash.
- **Run cold water** over the area.

What are the first aid steps for a burn?

Choking

▲ This is the universal choking sign. Make this sign if you are choking.

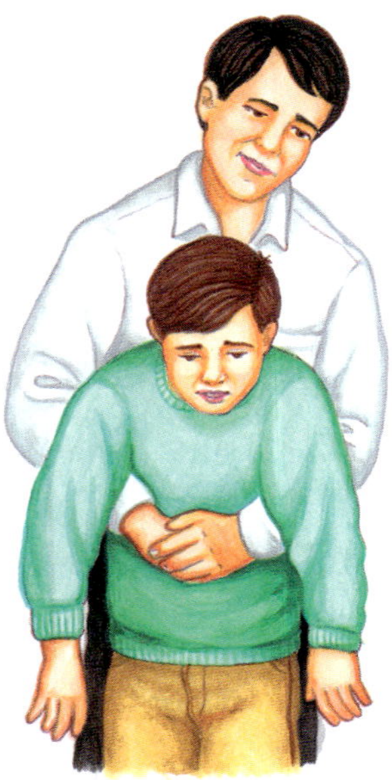

▲ Stand behind the person who is choking. Use quick, upward pushes into the stomach, just above the navel.

Sometimes a person may choke while eating. Chew your food slowly. Don't talk while you are chewing and swallowing. Remain seated. Don't eat and run. Be careful when eating certain foods that can get stuck in your throat, such as hot dogs and candies.

If someone is choking, follow these steps when giving first aid. You should be trained to do this before you try it.

1. **Ask the person, "Are you choking?"** The person might nod or make the sign for choking.
2. **Stand behind the person.**
3. **Wrap your arms around the person's waist.**
4. **Make a fist with one hand and grasp your fist with the other.** Press into the person's stomach just above their navel. Use quick, upward pushes. Don't press so hard that you break or injure the person's ribs.
5. **Give pushes until the food comes out.** Tell someone to dial 9-1-1 if the person does not stop choking.

What is the sign for someone who is choking?

Filling a First Aid Kit

A **first aid kit** is a collection of supplies used to care for a person who is injured or ill. Supplies include:

- gauze pads
- adhesive bandages
- towelettes
- scissors
- tweezers
- adhesive tape
- gauze roller bandages
- directions for emergency help

ACTIVITY LIFE SKILLS

CRITICAL THINKING

Be a Health Advocate

1. **Choose a healthful action to communicate.** What is the healthful action? You want to teach people about first aid.
2. **Collect information about the action.** Use your textbook to find information about first aid for cuts and scrapes.
3. **Decide how to communicate this information.** You might design a first aid brochure. Have one page for each type of first aid.
4. **Communicate your message to others.** Present your brochure to the class.

LESSON REVIEW

Review Concepts

1. **Explain** why you should not touch someone's blood while giving first aid.
2. **List** the first aid steps for a bruise.
3. **Name** four items often found in a first aid kit.

Critical Thinking

4. **Apply** Draw a picture of how to give first aid to a person who is choking. Use words to describe what to do.
5. LIFE SKILLS **Be a Health Advocate** You find out that only a few of the people you know have first aid kits at home. How can you get more of them to have first aid kits?

CHAPTER 6 REVIEW

accident, *C43*

emergency, *C69*

first aid kit, *C77*

personal flotation device, *C52*

thunderstorm, *C56*

violence, *C61*

Use Vocabulary

Choose the correct term from the list to complete each sentence.

1. Supplies that are needed to give first aid, such as gauze pads and adhesive bandages, are found in a(n) __?__.

2. Someone is having a hard time breathing. You should call for help quickly. This situation is a(n) __?__.

3. Never go near a stranger. A stranger could harm you. Harm done to yourself, others, or property is called __?__.

4. In order to protect yourself during a(n) __?__, which is a storm with thunder and lightning, keep away from water.

5. If you are riding in a boat, you should always wear an object that helps you stay afloat in water called a(n) __?__.

6. Something that is not supposed to happen is called a(n) __?__.

Review Concepts

Answer each question in complete sentences.

7. Name a hazard that might make you fall at home. Tell how you can keep safe in this example.

8. Explain how you can stay safe on a playground.

9. Describe the safe way to wait for a school bus.

10. Explain what to do if someone touches you in a way that is wrong.

11. Explain why gangs can be dangerous.

Reading Comprehension

Answer each question in complete sentences.

- **Have your parents check smoke alarm batteries.** Smoke alarm batteries should be checked at least twice a year.
- **Call for help if you see a fire.** Dial 9-1-1.

12. Tell who you should ask to check smoke alarm batteries.

13. Identify how many times smoke alarm batteries should be checked.

14. Name the number you should dial if you see a fire.

Critical Thinking/Problem Solving

Answer each question in complete sentences.

Analyze Concepts

15. You and your parents decide to ride bicycles down the road. Tell how you would signal a left turn.

16. You are planning to give a talk to first graders about ways to prevent choking. What information will you include in your speech?

17. During a game of street hockey, your friend falls and hurts his knee. His knee turns black and blue. Decide what type of injury your friend has. List the first aid steps for treating it.

18. When you get home from school, your parents are busy cleaning the attic. Describe what you should do to keep yourself safe if someone comes to the door.

19. At school, your class practiced what to do in case of a tornado. Describe the practice session.

20. What should you do if you get lost at a shopping mall?

Practice Life Skills

21. **Set Health Goals** Set a health goal to follow safety rules while riding in a car. Explain how this might affect your health.

22. **Make Responsible Decisions** You see a weapon on the playground at school. What should you do? Identify the responsible decision.

Read Graphics

The table shows first aid steps for a cut and a rash. Use the table to answer questions 23–24.

First Aid for Cuts

Injury	Steps
Cut	• Clean the cut with soap and water. • Cover the cut with a clean bandage.
Rash	• Run cold water over the area. • Do not scratch or rub the rash.

23. Describe how to clean a cut.

24. Identify what to run over a rash.

UNIT C ACTIVITIES AND PROJECTS

Effective Communication

Write a Skit

Write a skit. Two friends are on opposite football teams. One friend begins yelling and pushes the other friend. Include how the two friends can be good sports towards one another.

Self-Directed Learning

Follow Directions

Write the directions from your shampoo bottle on the left side of a two-column chart. Write how you wash your hair on the right. Compare both sides. Correct any steps that do not match.

Critical Thinking and Problem Solving

Electrical Safety

You want to plug your television, alarm clock, and lamp into one outlet. Explain why this is an unsafe situation. Decide which ones you could do without.

Responsible Citizenship

Design a Sign

Design signs to hang up outside on the playground at your school. The signs should have safety rules that everyone should follow in order to stay safe.

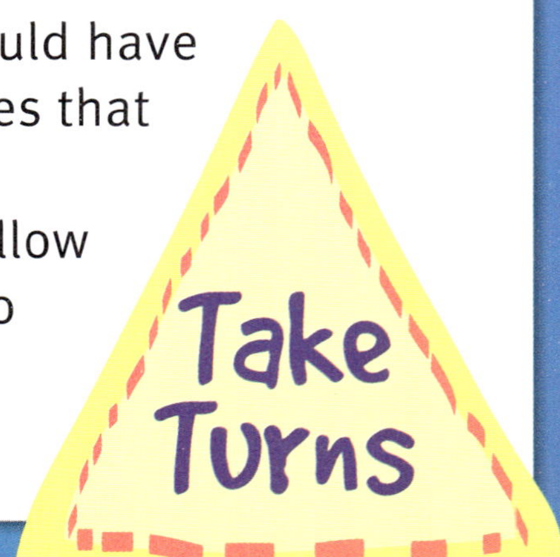

UNIT D

Drugs and Disease Prevention

CHAPTER 7

Alcohol, Tobacco, and Other Drugs

(bc)©Comstock/Alamy

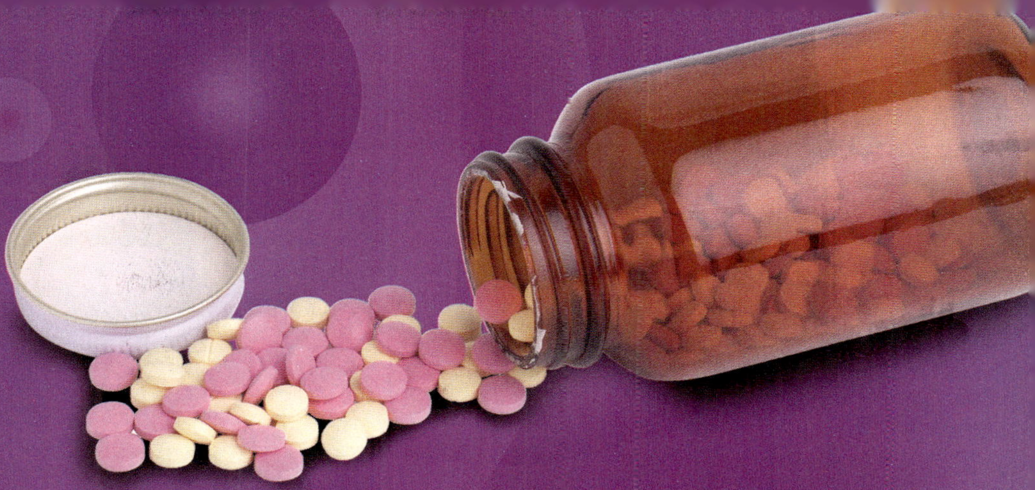

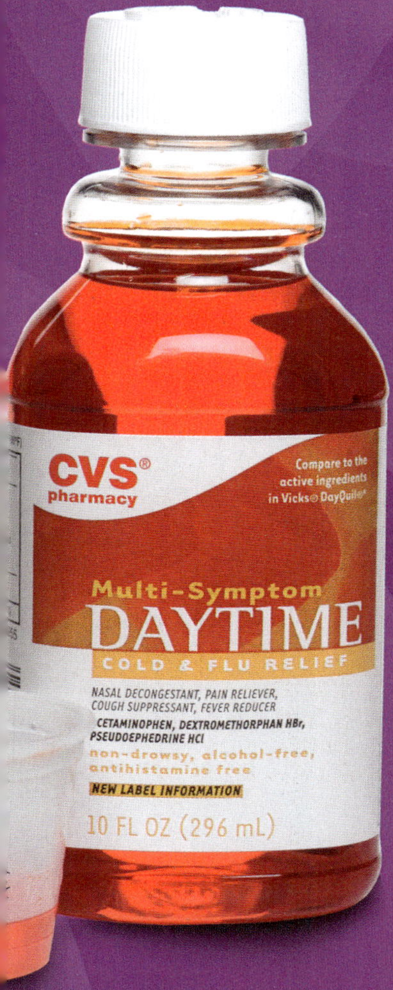

What Do You Know?

When people are ill, taking drugs can help them feel better. These drugs are medicine. What are some medicines that you know about? Put a check on the line if a term names a medicine.

_?_ honey
_?_ cough syrup
_?_ vaccination shots
_?_ chicken soup
_?_ cold tablets

Did you check cough syrup, vaccination shots, and cold tablets? Great! Read **Alcohol, Tobacco, and Other Drugs** to learn more.

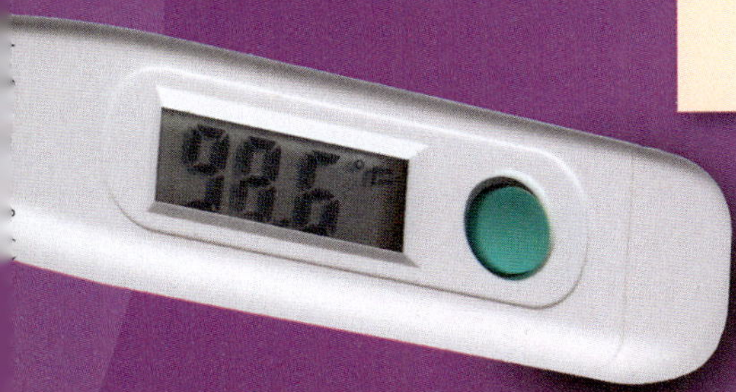

LESSON 1

Using Medicines Safely

You will learn . . .

- ways medicines can help health.
- how prescription and OTC medicines differ.
- what safety rules to follow for using medicines.

Jacob is feeling ill. Sometimes a doctor will give a person medicine to help him or her get well. Sometimes a parent or guardian will give medicine to a child. Jacob's mother is giving him some medicine to help him feel better.

Vocabulary

- **drug**, *D5*
- **medicine**, *D5*
- **prescription medicine**, *D6*
- **over-the-counter (OTC) medicine**, *D6*
- **drug misuse**, *D9*
- **drug abuse**, *D9*

Medicine and You

The adult in the photo is giving her child cough medicine. Cough medicine is a drug. A **drug** is a substance, other than food, that changes how your mind or body works. The parent is giving his daughter the medicine to help her stop coughing.

Medicines are drugs used to treat, prevent, or cure an illness or injury. Medicines can be helpful when used safely. Medicines can relieve a cough, heal burned skin, stop a runny nose, and prevent disease. Medicines come in many forms.

- **Tablets** These are swallowed with water.
- **Creams** Creams, such as ointments or lotions, are placed on your skin to clear up rashes.
- **Spray** Liquid medicine can be sprayed into your nose or mouth or onto your skin.
- **Injections** Medicines can be put into your body using a needle.

Medicines can be bought at drugstores. They are legal drugs. Legal drugs are safe when used as directed. Some drugs are illegal. They are not safe to use. Illegal drugs can harm your body. Say "no" to illegal drugs.

How do medicines help people?

ACTIVITY

Con$umer Wi$e

Comparing Prices

Go with your parent or guardian to a drugstore. Find two bottles of cough syrup. One bottle should be a brand name. The other bottle should be generic. Compare prices of each bottle. Which cough syrup is cheaper? How much cheaper? Which one would you buy?

▲ **Medicines can help you feel better when they are used correctly.**

Two Kinds of Medicines

CAREERS

Pharmacist

A pharmacist has a bachelor's degree and an advanced degree. He or she then has to pass a test to receive a license. A licensed pharmacist fills prescriptions written by doctors. Pharmacists work in drugstores, hospitals, and grocery stores. They also work for companies that make drugs.

A doctor might give you medicine when you are ill. He or she writes you an order for a prescription (prih•SKRIP•shuhn). **Prescription medicine** is medicine that your doctor writes an order for. Your parents or guardian can buy it at a drugstore. A pharmacist (FAHR•muh•sist) can fill the doctor's order for the medicine. No one else should take your prescribed medicine but you.

Over-the-counter (OTC) medicines are medicines adults can buy without a doctor's written order. These medicines may be used to treat a headache, upset stomach, cold, or body pain. Your parent or guardian will read the directions on the bottle. The directions tell the safe way to use the drug.

Name the two kinds of medicines.

Some OTC Medicines

Drug	How It Helps Health
Cough Medicine	Helps control coughs
Antibiotic (ant•ee•bigh•OT•tik) Ointment	Helps kill germs in cuts and scrapes
Allergy (AL•luhr•jee) Medicine	Helps relieve itchy eyes and runny noses

Medicine Labels

Medicines have labels. Each label has important information you should know. Read all medicine labels with your parents or guardian. Study the labels on this page.

BEST DRUG STORE Rx
150 SUNSHINE LANE NEW BOSTON, NH 03070

MR. PAUL SCOTT
1030 Hillcrest Ave
New Boston, NH 03070

TAKE 1 TABLET
EVERY 8 HOURS
Penicillin 200 mg
EXP: 7/30
Dr. Mark Smith

Directions Your parents or guardian will follow the directions on the prescription correctly. Never take medicine from anyone but your parents or guardian. Never take it on your own.

Dosage Always take the correct amount of medicine.

Name Never take prescription medicine that is prescribed for someone else. Be sure your name is on the label.

OTC Labels

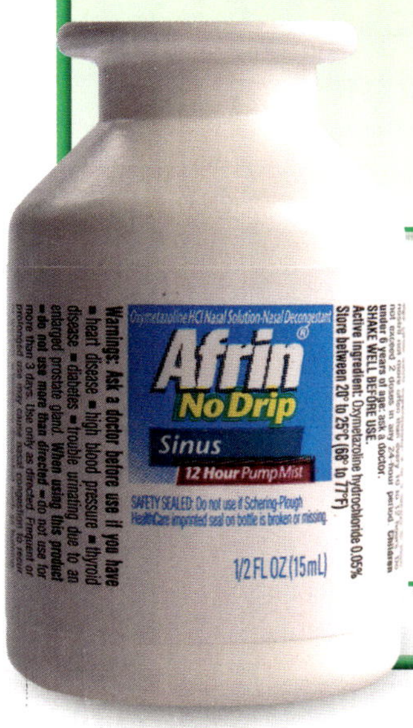

Warnings Your parents or guardian will read the warning label on the back of OTC medicines. The label tells who should not take the medicine. Some people are allergic to medicine.

Expiration Date Never take medicine that has passed the expiration date. An adult should get rid of expired medicine in a safe way.

Using Medicine Safely

Never take medicine unless given to you by a responsible adult such as your parent or guardian or your school nurse. It also is important to follow safety rules when taking medicine.

- Follow the directions on the label.
- Take all the medicine that a doctor prescribes. Do not use medicine that has passed the expiration date.
- Check the safety seal on OTC medicines before buying. Make sure it is not broken. Do not buy OTC medicines if the seal is broken.
- Tell an adult if you have a side effect from taking medicine. A *side effect* is an unwanted feeling or illness that occurs after taking medicine. Side effects may include sleepiness, a rash, itching, an upset stomach, or dizziness.
- Keep medicine in a safe place, away from young children.

Name the person who should give you your medicine.

Unsafe Use of Drugs

Drug misuse is the unsafe use of medicine taken by accident. For example, a person can take 3 teaspoons of medicine instead of 2 by mistake. You can prevent drug misuse by following the safety rules for taking medicines.

Drug abuse is the unsafe use of a medicine or an illegal drug on purpose. *Illegal* means against the law. For example, a person may have a prescription drug for headaches. The label on the drug reads, "Take one tablet a day." The person decides to take two or more tablets a day. This is drug abuse.

ACTIVITY

LIFE SKILLS

CRITICAL THINKING

Be a Health Advocate

Everyone should follow the safety rules for taking medicine. What can you do to remind others?

1. **Choose a healthful action to communicate.** What safety rules will you share with others?
2. **Collect information about the action.** Your text has information on safety rules for taking medicine. You could also learn more by doing research in the library.
3. **Decide how to communicate this information.** Will you record a special announcement? Or will you make a poster or a calendar?
4. **Communicate your message to others.** You might share your announcement over the PA. Ask your principal for permission.

LESSON REVIEW

Review Concepts

1. **Explain** ways medicines can help health.
2. **Compare** the two kinds of medicines.
3. **List** three safety rules for using medicine.

Critical Thinking

4. **Apply** You have an upset stomach. Your friend offers you his or her prescription drug. What should you do? Explain your answer.
5. LIFE SKILLS **Be a Health Advocate** What steps can you follow to be a health advocate for the safe use of medicines?

LESSON 2

Alcohol

You will learn . . .

- ways that alcohol affects physical health.
- ways that alcohol affects how a person thinks, feels, and acts.
- ways to say "no" to drinking alcohol.

Vocabulary

- **alcohol**, *D11*

This boy and girl are having fun. They have good health. Drinking alcohol can harm health. Say "no" to drinking alcohol.

Alcohol Affects Physical Health

Alcohol is a drug that slows down the mind and the body. It is found in some drinks. When a person drinks alcohol, it quickly enters the blood. The blood carries the alcohol to all parts of the body. The drawing below shows the different parts of the body affected by alcohol.

What is alcohol?

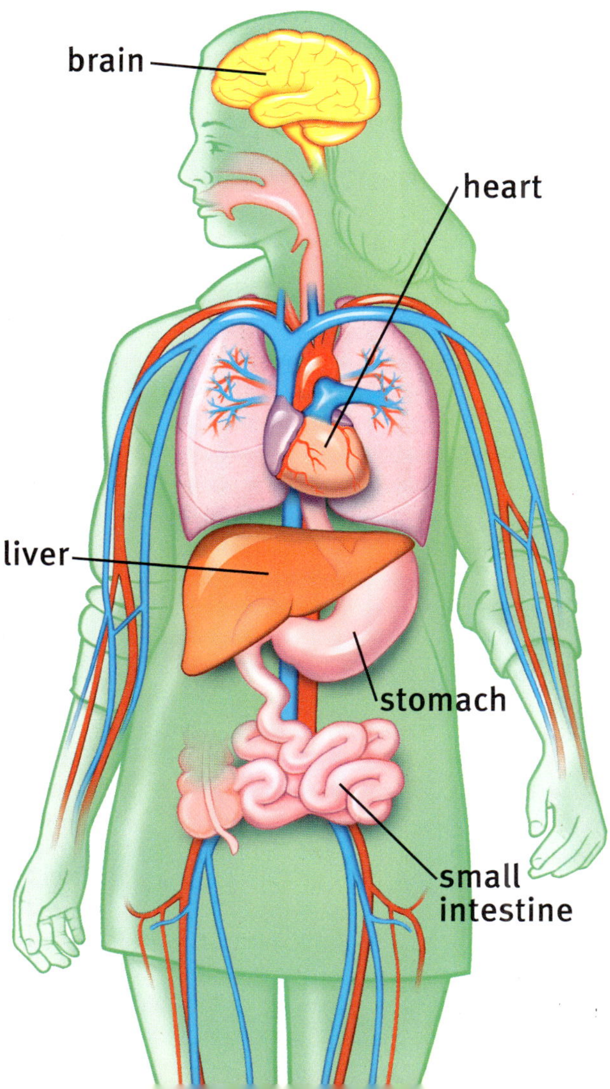

How Alcohol Affects the Body

- **Brain** Alcohol slows down the way the brain works. A person can't think clearly or react correctly to messages the body sends to the brain.
- **Heart** Heavy drinking can cause heart failure.
- **Liver** Too much alcohol can harm the liver.
- **Stomach** Too much alcohol can harm stomach cells.
- **Small Intestine** Too much alcohol can harm the small intestine and cause digestion problems.
- **Senses** Alcohol dulls the senses of sight, hearing, smell, touch, and taste.

Alcohol Changes the Way People Act

ACTIVITY

Physical Education LINK

IT Alert

Test your alertness. Play this game with a group of six or more. Choose one player to be IT. The rest of the group will stand in a circle around IT. Give every player a number, including IT. Call out two numbers. The players whose numbers are called must try to change places. IT tries to move into an open space before either player gets there. The player who is left out will be the new IT.

The brain is the control center of the body. A person who drinks alcohol might not be in control of what he or she does. The person might not be alert, or ready to act. When drinking alcohol, people often

- make wrong decisions.
- say things that hurt other people's feelings.
- forget things they already have learned.
- become angry and get into fights when there is a conflict.
- are unable to walk and talk properly because they cannot control their muscles.
- become sad.

▼ **Drinking alcohol and driving is against the law. It can lead to car crashes.**

Drinking Alcohol Can Harm Others

People who drink alcohol can harm themselves and others. Here are three ways others could be harmed.

- **Drinking alcohol can cause accidents.** People can do dangerous things like driving after drinking. Drinking and then driving causes accidents.
- **Drinking alcohol can cause crime and violence.** People who drink alcohol aren't able to make responsible decisions. They may show their feelings in ways that harm others.
- **Drinking alcohol can cause problems in a family.** Family members who drink often are not as dependable or responsible as those who don't drink. They might forget a family function.

What are three ways that drinking alcohol can change the way people act?

ACTIVITY

Science LINK

Learning with Balloons!

You will need a red and blue balloon for this activity. The red balloon represents a healthy stomach. The blue balloon represents a stomach damaged by alcohol. With your teacher, use a paper clip to poke holes through the blue balloon. Now fill both balloons with water over the sink. Describe what happened. How is the blue balloon different from the red one?

MADD stands for *Mothers Against Drunk Driving*. MADD is an organization that educates young people and adults about the dangers of alcohol. One of MADD's goals is to stop drunk driving.

Say "No" to Alcohol

You can use resistance skills if someone pressures you to drink alcohol.

1. **Look at the person. Say "no" in a firm voice.**
2. **Give reasons for saying "no."**
   - I want to protect my health and keep my mind clear.
   - I want to be safe and avoid accidents.
   - I want to follow rules and laws.
   - I want to show respect in my relationships.
   - I want to follow my family guidelines.
   - I want to be alcohol free and show good character.
3. **Match your behavior with your words.** Walk away if someone wants you to drink.
4. **Ask an adult for help if you need it.** Tell your parents or guardians what happened. They will help you stick to your no.

▲ **Say "no" to alcohol, and stay safe.**

What are two reasons to say "no" to alcohol?

ACTIVITY

LIFE SKILLS

CRITICAL THINKING

Use Communication Skills

Role-play a conversation between Alice and Tina. Alice wants to make a poster to teach students about the harmful effects of alcohol. Tina does not feel that sharing health facts is important.

1. **Choose the best way to communicate.** Alice will remember to use good communication skills.
2. **Send a clear message. Be polite.** What should Alice say to Tina? How can she tell her that sharing healthful facts is important?
3. **Listen to each other.** How can Tina let Alice know that she is listening? How can Alice let Tina know that she is listening?
4. **Make sure you understand each other.** How should Tina let Alice know that she understands her?

LESSON REVIEW

Review Concepts

1. **List** three ways that alcohol harms physical health.
2. **Describe** how alcohol affects relationships.
3. **List** the steps to follow to say "no" if someone asks you to drink alcohol.

Critical Thinking

4. **Apply** Create a poster that will tell people how drinking alcohol can harm others. What should the poster say?
5. LIFE SKILLS **Use Communication Skills** Your friend's teenage brother wants to know if you want a ride home from school. You and your friend see an open can of beer in his car. What should you do?

LESSON 3

Tobacco

You will learn . . .

- how smoking and smokeless tobacco harm health.
- ways secondhand tobacco smoke harms health.
- ways ads try to get people to use tobacco.
- ways to say "no" to tobacco use.

Vocabulary

- **tobacco**, *D17*
- **nicotine**, *D17*
- **dependence**, *D17*
- **secondhand smoke**, *D18*

Cigarettes are a tobacco product. They contain harmful chemicals. There are over 4,000 harmful chemicals in tobacco smoke. Many buildings have no smoking signs. This helps to keep the environment in the building healthy.

Tobacco Has Harmful Chemicals

Tobacco is a plant. Tobacco is found in cigarettes, cigars, pipe tobacco, and chewing tobacco. It contains many harmful chemicals, like nicotine.

Nicotine (NI•kuh•TEEN) is a drug found in tobacco. It speeds up your body. It makes your heart beat faster, for example. People who use tobacco can develop a dependence on nicotine. **Dependence** is a strong need for something. Once someone has nicotine dependence, it becomes harder to quit using tobacco. Over time, tobacco use can harm many parts of your body. Look at the diagram below.

What does having a dependence on a drug mean?

Ways Chemicals in Tobacco Can Harm Health

- **Face** Causes wrinkles and makes you look older.
- **Mouth** Causes gum disease and cancer.
- **Throat** Can cause throat cancer.
- **Lungs** Can cause frequent colds and lung disease.
- **Heart** Can cause high blood pressure and heart disease.
- **Stomach** Can cause stomach cancer and ulcers.

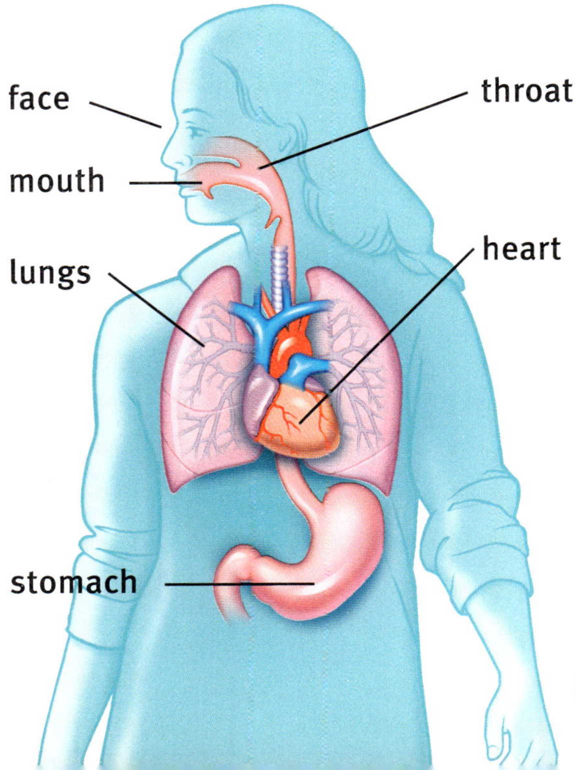

Secondhand Smoke

Have you ever been in a room where people were smoking? If you have, then you were breathing secondhand smoke. **Secondhand smoke** is the smoke that a smoker breathes out. It also is the smoke from a burning cigarette, cigar, or pipe. Secondhand smoke can make your eyes water and itch. It can make you cough. Over time it can even cause heart disease, lung cancer, and breathing problems.

You and your family should avoid secondhand smoke whenever possible. You can ask to sit in the nonsmoking section in restaurants, for example.

How can secondhand smoke harm you?

Michelle Bachen

Tobacco Ads

Tobacco companies advertise their products in magazines, on billboards, at sporting events, and at concerts. The companies advertise to get people to buy cigarettes, cigars, and chewing tobacco. But the law says tobacco ads must warn people that tobacco harms health. The warnings say that cigarette smoke has harmful gases that can cause disease.

Explain why a tobacco ad might show a group of people smoking.

▼ **The pictures below show how tobacco companies try to sell products. What messages are the ads sending to consumers?**

ACTIVITY

Art LINK

Make an Anti-Tobacco Ad

Make an anti-smoking ad. Cut out pictures of people from magazines. Arrange them on poster board. Use speech bubbles. Have people say "no" to tobacco. Is your ad convincing? Would someone want to smoke after looking at your ad?

How Tobacco Companies Advertise

▶ **You will look as good as the model if you smoke.**

▶ **You will be successful if you smoke.**

◀ **You will have fun if you smoke.**

▲ **Other people will like you if you smoke.**

(l to r) ©PhotoAlto; ExactoStock/SuperStock; ©PhotoAlto; Purestock/SuperStock

Use Resistance Skills

1. Look at the person. Say "no" in a firm voice.
2. Give reasons for saying "no."
3. Match your behavior to your words.
4. Ask an adult for help if you need it.

Guidelines for Making Responsible Decisions™

to say no to smoking:

- Is it healthful? No.
- Is it safe? No.
- Does it follow rules and laws? No.
- Does it show respect for me and others? No
- Does it follow family guidelines? No.
- Does it show good character? No.

Say "No" To Tobacco

Remember one word when someone offers you tobacco. The word is "no." You know that using tobacco is harmful to your health. It is also illegal for young people under the age of 18 to purchase tobacco.

Think of the diagram of the human body in this lesson. Think of all the parts of your body that tobacco can harm. Decide now that you will not use tobacco.

Sometimes a friend or an older student might offer you a cigarette. Tell him or her that you will not use tobacco. You want to live tobacco-free. Stand up for good health and help your friends decide to say "no."

▼ **These students have healthy smiles. They say "no" to smoking.**

ACTIVITY
LIFE SKILLS
CRITICAL THINKING

Access Health Facts, Products, and Services

1. **Identify when you might need health facts, products, and services.** How might learning about the harmful effects of tobacco help you?
2. **Identify where you might find health facts, products, and services.** This textbook has lots of information on the harmful effects of tobacco. What other resources might help you?
3. **Find the health facts, products, and services you need.** Gather information about the harmful effects of smoking and make flash cards.
4. **Evaluate the health facts, products, and services.** Ask the school nurse or your family doctor for other facts that you should include on your cards.

What should you do if someone offers you tobacco?

LESSON REVIEW

Review Concepts

1. **Explain** what nicotine is and how it harms the body.
2. **Define** secondhand smoke.
3. **Tell** what you can say to a person who offers you a cigarette.

Critical Thinking

4. **Examine** Discuss what you should look for in tobacco advertisements.
5. LIFE SKILLS **Access Health Facts, Products, and Services** A family member smokes. He or she gets upset when others suggest quitting. What should you do?

Learning LIFE SKILLS

CRITICAL THINKING

Use Resistance Skills

Problem Some older students are pressuring you to take just one puff from a cigarette. How can you use resistance skills to say "no"?

Solution Saying "no" to smoking a cigarette is a responsible decision. Use the steps on the next page to practice how to say "no."

Foldables™ To Learn Life Skills

Learn This Life Skill

Follow these steps to use resistance skills. The Foldables™ can help you.

1 Look at the person. Say "no" in a firm voice.

Make sure that the person pressuring you to smoke understands that you really mean "no."

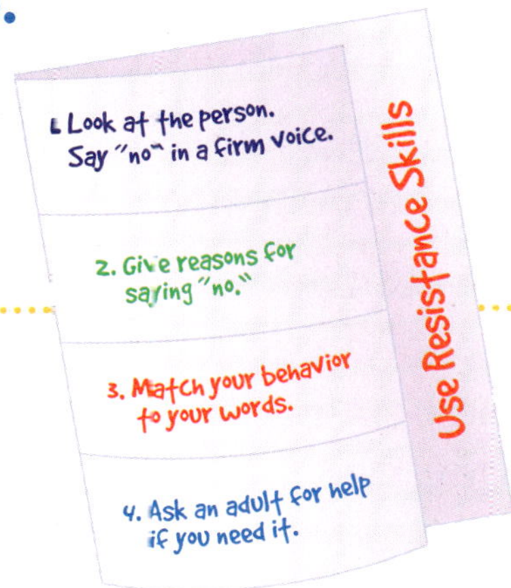

2 Give reasons for saying "no."

Tell the person that you want to protect your health. What other reasons might you give?

3 Match your behavior to your words.

Walk away. Try to avoid going to places where people will pressure you to smoke.

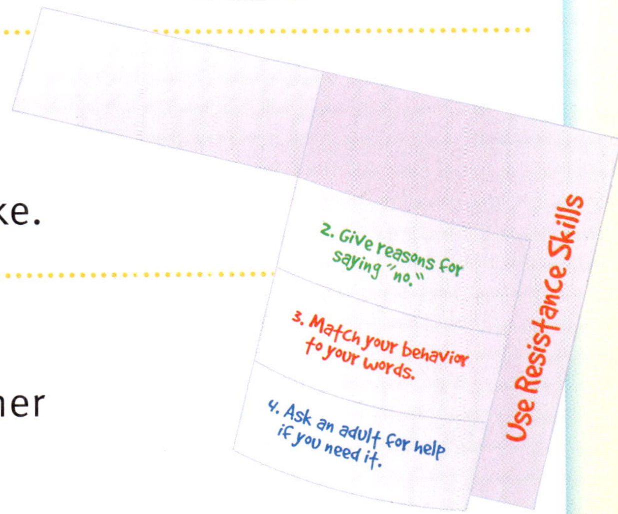

4 Ask an adult for help if you need it.

Talk to your parents, guardian, or another responsible adult. Get their advice.

Practice This Life Skill

With your classmates, role-play the situation above. Use the steps for using resistance skills.

LESSON 4

Other Drugs

You will learn . . .

- ways caffeine can harm health.
- how illegal drug use can harm health.
- how to say "no" to illegal drugs and the benefits of a drug-free lifestyle.
- how to get help for drug abuse.

Vocabulary

- **caffeine**, *D25*
- **marijuana**, *D26*
- **inhalant**, *D26*
- **cocaine**, *D27*
- **crack**, *D27*
- **ecstasy**, *D27*

Drugs can be found in foods and drinks, like chocolate and soft drinks. You can choose to limit or avoid these foods and drinks.

Caffeine Affects the Mind and Body

Caffeine (kaf•EEN) is a drug that speeds up body actions. It can be found in some colas, coffee, tea, and chocolate. Caffeine changes how your mind or body works. Here are some examples of what caffeine can do to you:

- Your heart beats faster.
- You might feel nervous or shaky.
- You might feel more awake.
- You might get an upset stomach.

Some products say "decaffeinated," "no caffeine," or "caffeine-free." Choose these products. Or choose water, milk, or fruit juice. These drinks provide nutrients that help your body grow. They do not contain caffeine.

Why is caffeine a drug?

ACTIVITY

Health Online

Caffeine: Then and Now

Research the history of caffeine. Use the e-Journal writing tool to write your report.

▶ **Look for the words "caffeine-free" when choosing a cola to drink.**

Harmful Effects of Illegal Drugs

Some drugs are illegal. This means that it is against the law to buy, sell, or use these drugs. It is never safe to use an illegal drug.

Say "No" to Marijuana

Marijuana (mayr•uh•WAH•nuh) is an illegal drug that contains harmful chemicals. People who use marijuana get confused and find it hard to think. Smoking marijuana can harm the lungs and heart, just as smoking cigarettes can.

Say "No" to Inhalants

An **inhalant** (in•HAY•luhnt) is a chemical that is breathed into the lungs. Some inhalants are medicines. A doctor writes an order for these medicines.

▼ It is important to read warning labels when using products that give off fumes.

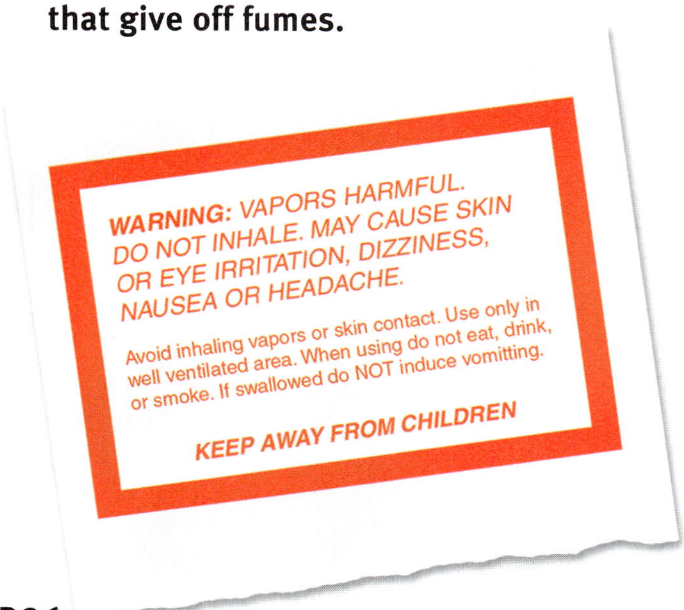

Other inhalants are fumes that can come from materials like glue, spray paint, and gasoline.

Never breathe in these fumes on purpose. They can cause headaches, nosebleeds, brain damage, and death. Always read the warnings on the labels. Have your parents or guardian read the warnings with you. Open a window to get fresh air.

Other Illegal Drugs

Cocaine (koh•KAYN) is an illegal drug that is made from the coca plant. It is a white powder. People who use cocaine inhale or inject it. Even young people can have heart attacks and die from using this drug.

Crack is an illegal drug that is a form of cocaine. This illegal drug is smoked. It affects the body in seconds.

Another illegal drug is ecstasy (EK•stuh•see). **Ecstasy** speeds up what happens inside the body. This drug is swallowed as a tablet or capsule. It also can be inhaled as a powder. Ecstasy can cause fainting, confusion, and brain damage.

What is an illegal drug?

▼ Choose a drug-free life. It is one of the most healthful choices you can make.

ACTIVITY

BUILD Character

Pledge to Be Drug-Free

Citizenship Write "I Pledge to Be Drug-Free" at the top of a sheet of paper. Then write the benefits of being drug-free. Sign your name and write the date at the bottom.

To be drug-free is to:

- Keep my body healthy
- Follow laws
- Show good character
- Keep friendships
- Get good grades
- Stay out of fights
- Keep my mind alert
- Avoid accidents and injuries

Say "No" to Drugs

You can use resistance skills if someone pressures you to use drugs.

1. **Look at the person. Say "no" in a firm voice.**
2. **Give reasons for saying "no."**
   - I want to protect my health and keep my mind clear.
   - I want to follow rules and laws.
   - I want to show good character.
3. **Match your behavior with your words.** Walk away if someone wants you to use drugs.
4. **Ask an adult for help if you need it.** Tell your parents or guardian what happened. They will help you stick to your no.

What adults can you turn to for help when you say "no" to drugs?

▶ **Always say "no" to drugs.**

Getting Help for Drug Abuse

A person can get "hooked" on some drugs very quickly. It can be hard for this person to stop using drugs.

A person who can't stop taking a drug needs help. The person can talk to a doctor or counselor about different programs that help people to recover. Tell your parents, guardian, or another responsible adult if you know someone who is abusing drugs.

ACTIVITY LIFE SKILLS

CRITICAL THINKING

Make Responsible Decisions

You can choose a cola drink, milk, grape juice, or bottled water. Which should you choose?

1. **Identify your choices. Check them out with your parent or trusted adult.** Make a four-column chart. Place each drink choice at the top of a column.
2. **Evaluate each choice. Use the *Guidelines for Making Responsible Decisions*™.** Write answers to the guideline questions in the chart.
3. **Identify the responsible decision. Check this out with your parent or a trusted adult.** Support each decision using answers from the chart.
4. **Evaluate your decision.** Discuss the possible outcomes of the decision.

LESSON REVIEW

Review Concepts

1. **Discuss** two ways that caffeine can affect the mind and body.
2. **Explain** why marijuana and cocaine are illegal. Give two ways they can harm health.
3. **Identify** what you should do if you use a product that produces fumes.

Critical Thinking

4. **Identify** Name two benefits of a drug-free lifestyle.
5. LIFE SKILLS **Make Responsible Decisions** A classmate pressures you to smoke marijuana with him. What is the responsible decision for you to make?

CHAPTER 7 REVIEW

Use Vocabulary

alcohol, *D11*
caffeine, *D25*
dependence, *D17*
drug, *D5*
drug misuse, *D9*
nicotine, *D17*
over-the-counter medicine (OTC), *D6*

Choose the correct term from the list to complete each sentence.

1. A drug that speeds up body actions is called __?__.
2. When a substance other than food changes how your mind or body works, it is called a __?__.
3. Medicine that adults can buy without a doctor's written order is called a(n) __?__.
4. A drug found in tobacco that speeds up the body is called __?__.
5. A drug that slows down your mind and body is called __?__.
6. The unsafe use of medicine taken by accident is called __?__.
7. A strong need for something is called __?__.

Review Concepts

Answer each question in complete sentences.

8. What are three forms in which medicine might be given to you?
9. List two ways that alcohol might change the way a person feels.
10. Name the one thing that every tobacco advertisement must show.
11. Describe what happens when a person gets "hooked" on drugs.
12. Tell two places that OTC drugs can be purchased by an adult.

Reading Comprehension

Answer each question in complete sentences.

An inhalant (in•HAY•luhnt) is a chemical that is breathed into the lungs. Some inhalants are medicine. They can be used if a doctor prescribes them.

Other inhalants are fumes that can come from glue, spray paint and gasoline.

13. Define an inhalant.
14. Identify three types of inhalants.
15. Explain when an inhalant might be used to help health.

Critical Thinking/Problem Solving

Answer each question in complete sentences.

Analyze Concepts

16. Two children in the same family have ear aches. Why would each child receive a different prescription?

17. Discuss why it is important to say "no" to alcohol.

18. While visiting a hospital, an older patient who smoked for 30 years complains about shortness of reath. What might have harmed patient's health? How?

19. re at a baseball game with o iend. The people in front ho re smoking. Summarize your ndhand smoke affects .

Practice Skills

20. **Use Resis Skills** You hear high school stud at school talking about ecstasy ey say it makes them feel good. t three reasons why you should ys say "no" to illegal drugs.

21. **Make Responsible isions** After taking a medicin or a headache, you begin to el sleepy. Your stomach be s to hurt. What responsible decision hould you make?

Read Graphics

The graph below plots the number of towns that banned smoking in public places from 1985 to 2003. Use the graph to answer the questions.

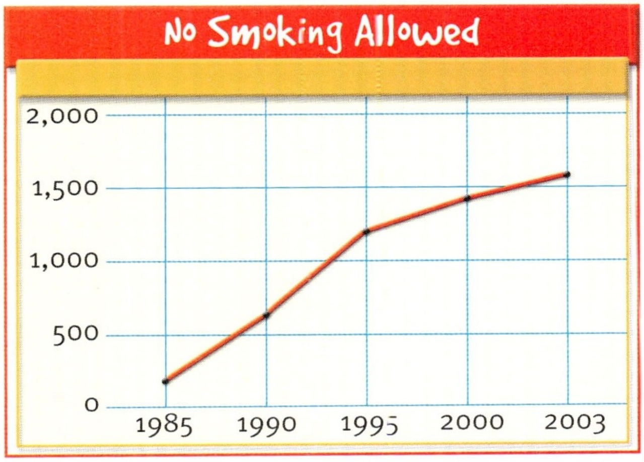

22. About how many towns in 1990 did not allow smoking inside public places?

23. About how many towns in 2003 did not allow smoking inside public places?

24. What does this graph tell you about smoking in public places?

CHAPTER 8

Communicable and Chronic Diseases

What Do You Know?

Germs are all around you. They spread in many different ways. Read the following statements. Answer **yes** or **no**.

_?_ I wash my hands after using the restroom.
_?_ I put insect spray on my skin when I play outdoors.
_?_ I wash an apple before I eat it.

Did you answer **yes** to all the statements? There are many things you can do to keep germs from spreading. Read **Communicable and Chronic Diseases** to learn more about germs and the diseases they cause.

LESSON 1

Diseases That Spread

You will learn . . .

- what causes diseases.
- ways to prevent the spread of disease.
- how to treat flu, cold, and strep throat.

Vocabulary

- **communicable disease**, *D35*
- **germ**, *D35*
- **bacteria**, *D35*
- **virus**, *D35*
- **symptom**, *D38*
- **fever**, *D38*

Lisa likes to play outdoors. She puts on lotion before she goes outside to keep insects off her skin. She knows that some insects spread disease. A *disease* is an illness that keeps your body from working as it should.

What Causes Disease

A **communicable** (kuh•MYEW•ni•kuh•buhl) **disease** can be spread to people. These diseases include colds, flu, sore throat, and strep throat. Germs cause most diseases that spread. A **germ** is a tiny particle that can cause disease.

Kinds of Germs

There are two different kinds of germs. **Bacteria** (bak•TEER•ee•uh) are one-celled germs. Bacteria live in and on living and nonliving things. Bacteria reproduce by dividing in the middle of one cell to produce two cells. Different bacteria cause different *infections* and diseases. An *infection* is the growth of germs that cause harm to the body. Bacteria known as strep cause strep throat. Some skin infections are caused by staph, another kind of bacteria.

A **virus** (VIGH•ruhs) is a very tiny germ that reproduces inside a living cell. Viruses get into your body cells and make exact copies of themselves. Viruses cause colds, flu, and chicken pox.

What is a disease that can spread to people?

ACTIVITY

Science LINK

Examine Bacteria

Bacteria are small. You cannot see them without using a microscope. Scientists use microscopes to look at these tiny germs. Examine the photographs below. How are the bacteria shapes different?

▲ Bacteria can be seen only with a microscope.

ACTIVITY

On Your Own

FOR SCHOOL OR HOME

Stop Germs! Poster

Family members share living space. They also spend lots of time together. Germs can spread quickly in families. What are some ways family members can slow or prevent the spread of germs? Use poster board and markers to make a poster to show your ideas. Hang the poster where family members can see it.

How Germs Spread

Germs can spread in four main ways. If you know how germs spread, you can practice healthful habits to prevent them from spreading.

The Air A sneeze or cough can send liquid drops into the air. These liquid drops contain germs. If someone close to you coughs or sneezes, you could inhale these liquid drops and become ill. The liquid drops could also enter your body through a break in your skin. Cover your nose and mouth when you cough or sneeze. This will prevent germs from spreading through the air. If you use your hands to cover your cough or sneeze, wash them. This will prevent you from spreading germs to people or objects you touch.

Contact with People and Objects People and objects sometimes carry germs. Germs spread when you touch people or objects. The germs pass into your body through your eyes, nose, or mouth. Keep your hands clean and away from your eyes, nose, and mouth. Always wash your hands after going to the restroom. This will prevent you from spreading germs to people or objects you touch.

Contact with Animals Germs get into your body through broken skin caused by an animal bite or scratch. Some germs from animals can spread. Do not pet strange animals. Have a parent, guardian, or responsible adult put sprays and lotions on your skin to help keep mosquitos and ticks off you.

Food and Water Germs live in lakes, ponds, and streams. Water from lakes, ponds, and streams must be treated with special chemicals. These chemicals make the water healthful to drink. Do not drink water that has not been treated.

Some foods you buy may have germs. Germs may be on fresh fruits and vegetables. Germs also can be in meats or eggs. The germs can be killed if the food is properly prepared and cooked.

Always wash your hands before you prepare or eat food. Do not prepare food for others when you are ill. Wash raw fruits and vegetables before you eat them. Do not eat undercooked hamburger or chicken. Do not eat foods that are made with raw eggs, like uncooked cookie dough. Wrap and refrigerate leftovers right away.

Why should you cover your mouth when you sneeze or cough?

Washing Hands Correctly

1. Wet your hands under warm water. Apply soap to both of your hands.
2. Constantly rub your hands together as you slowly count to 15.
3. Rinse your hands well. Dry with a paper towel.
4. Use the paper towel to turn off the water. Throw the paper towel away.

Flu, Colds, and Strep Throat

Do You Know

The average temperature of the human body is 98.6 degrees. If you have a fever, your body temperature could rise 1 to 5 degrees above your average temperature.

Symptoms of a disease tell you that you are ill. A **symptom** is a change in your body that is a sign of disease. Fever, cough, and runny nose are common symptoms of diseases such as the flu or a cold. **Fever** is a body temperature that is higher than normal.

Colds, flu, sore throats, and strep throat are common communicable diseases. Each disease has different symptoms.

Colds and Flu

Viruses cause colds and flu. A runny or stuffy nose, sneezing, coughing, and headache are symptoms of a cold and the flu. Fever is a symptom, too.

There is no cure for a cold or the flu. However, there are medicines that treat the symptoms of these diseases. Some medicines lower fever. Others may ease a cough. Getting rest and drinking lots of water and juice also help your body fight these diseases.

▲ **It is important to rest and drink lots of water and juice to help your body fight disease.**

What are common symptoms of disease?

Sore Throat

Both bacteria and viruses can cause a sore throat. There is no cure for a sore throat caused by viruses. Medicines like cough syrup and cough drops can help soothe a sore throat.

Strep Throat

Bacteria cause strep throat. Symptoms of strep throat are a sore throat, swollen glands, difficulty swallowing, and fever. Your throat may be very red with white patches.

If not treated, strep throat can lead to heart and kidney disease. Your doctor can do a test to tell you whether you have strep throat. The doctor will give you medicine to treat the disease.

ACTIVITY LIFE SKILLS CRITICAL THINKING

Resolve Conflicts

Role-play a conflict between Johnny and Rick. Johnny has a cold and a runny nose. He asks Rick for a bite of his fruit bar. Rick says "no."

1. **Stay calm.** To begin the role-play, Johnny acts upset at Rick. Rick reacts calmly.
2. **Talk about the conflict.** Rick explains to Johnny why he does not want to share his fruit bar. Johnny listens.
3. **List possible ways to settle the conflict.** Rick and Johnny use the *Guidelines for Making Responsible Decisions*™ to help settle the conflict. Rick and Johnny suggest ways to settle the conflict.
4. **Agree on a way to settle the conflict. You may need to ask a responsible adult for help.** Rick and Johnny agree on the best solution.

LESSON REVIEW

Review Concepts

1. **Name** two kinds of germs that cause disease. Name a disease that each germ causes.
2. **Describe** two ways to prevent the spread of disease.
3. **Discuss** symptoms of colds, flu, and strep throat.

Critical Thinking

4. **Analyze** Why is it important for strep throat to be detected and treated early?
5. LIFE SKILLS **Resolve Conflicts** A friend with a cold wants to drink from your cup. Describe how to resolve this conflict.

Fighting Disease

You will learn . . .

- ways the body keeps germs out.
- how the immune system fights disease.

Vocabulary

- **immune system**, *D41*
- **antibody**, *D41*
- **vaccine**, *D41*
- **medicine**, *D42*

You fall and cut your knee. Your parent or guardian washes the cut and puts cream on it. Next, he or she covers the cut with a bandage. Proper care of cuts and scrapes helps keep you safe from germs and fights disease.

How Your Body Keeps Out Germs

Your body has ways to protect you from germs. One way is through your immune system. The **immune** (i•MYEWN) **system** is the body system that protects the body from disease.

White Blood Cells

Your blood is made up of different types of cells. One type of cell is called a white blood cell. White blood cells help keep you well by destroying germs that get into your body.

Antibodies

Your immune system also uses antibodies to fight disease. An **antibody** (AN•ti•bod•ee) is a substance in your blood that helps the body kill germs. Your body makes a different antibody to help kill each kind of germ.

Your body makes antibodies two ways. First, it makes antibodies each time you get a disease caused by a germ. Second, your body makes antibodies when a doctor gives you a vaccine. A **vaccine** (vak•SEEN) is a substance made with dead or weak germs.

Why does your body make many kinds of antibodies?

Do You Know

The average person has 7,000 white blood cells in each milliliter of blood.

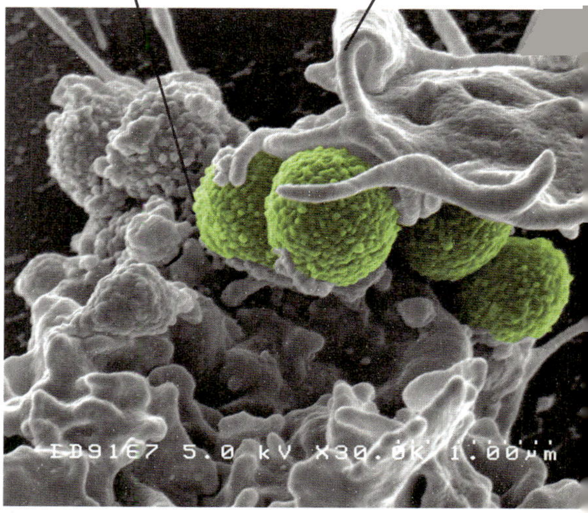

▲ White blood cells kill viruses and bacteria that enter your body.

Your Body Fights Disease

Your immune system fights disease. You can help your immune system work well. Tell an adult if you feel ill. He or she can help you get treatment or take you to a doctor. A doctor may give you medicine. **Medicine** (MED•uh•sin) is a drug used to treat an illness or injury. Treatment may include rest or drinking water or juice. Sometimes it includes taking medicine a doctor prescribes.

A doctor may treat diseases caused by bacteria with antibiotics. *Antibiotics* are a kind of medicine used to kill bacteria. Strep throat is treated with antibiotics. When a doctor gives you antibiotics, it is important to finish all of the medicine. Do not stop taking the medicine when you feel better. All the bacteria may not be killed. They might grow again. This can make you ill again.

Vaccines

A doctor may give you a vaccine to keep you from getting a disease. A vaccine is a substance made with dead or weak germs. When the vaccine enters your blood, your body makes antibodies to fight that germ. Different vaccines protect you from different germs.

What are antibiotics?

ACTIVITY

Science LINK

When Antibiotics Attack!

Collect a clear glass, cotton balls, and blue food coloring. Pretend the glass is your body. Fill it with water to represent blood. Place a cotton ball in the water to represent bacteria in your blood. To fight bacteria, a doctor can prescribe antibiotics. Put three drops of blue food coloring in the water to represent an antibiotic. Place a lid on the glass and shake it. The cotton ball turns blue. The bacteria have been killed by the antibiotic. How does this exercise reflect how antibiotics kill bacteria in your body?

ACTIVITY

LIFE SKILLS

CRITICAL THINKING

Manage Stress

Eva is going to the doctor. She is getting a vaccine. Eva is nervous. Work in a small group. Have one group member copy the questions below onto a sheet of chart paper. Each group member should answer the questions.

1. **Identify the signs of stress.** What are some signs of stress Eva may be feeling?
2. **Identify the cause of stress.** Besides going to the doctor, what may be causing Eva stress ?
3. **Do something about the cause of stress.** What can Eva do to manage her stress?
4. **Take action to reduce the harmful effects of stress.** What are some healthful behaviors Eva can practice each day to manage stress? Review each group member's answers to the questions. Write a plan for Eva that will help her manage stress. Present your group's plan to the class.

LESSON REVIEW

Review Concepts

1. **Describe** how your skin protects you from disease.
2. **Name** the type of blood cell your immune system uses to fight disease.
3. **Identify** three ways to help treat a disease caused by a virus.
4. **Explain** how your body uses a vaccine to fight germs.

Critical Thinking

5. **Infer** Sam had an ear infection two weeks ago. He took the antibiotics his doctor gave him just until he felt better. He now has another ear infection. Why might the infection be back?
6. LIFE SKILLS **Manage Stress** Your mom is feeling stress. She has been taking care of your ill sister. Discuss why your mom is feeling stress.

Learning LIFE SKILLS

CRITICAL THINKING

Practice Healthful Behaviors

Problem Your brother has the flu. Many children at your school also have the flu. You don't want to become ill. You begin thinking about ways you can avoid germs.

Solution How can you protect yourself from germs? You can choose healthful habits that prevent the spread of germs. Using the four steps on the next page can help.

Foldables™ To Learn Life Skills

Learn This Life Skill

Follow these steps to help you practice healthful behaviors that prevent the spread of germs. The Foldables™ can help you.

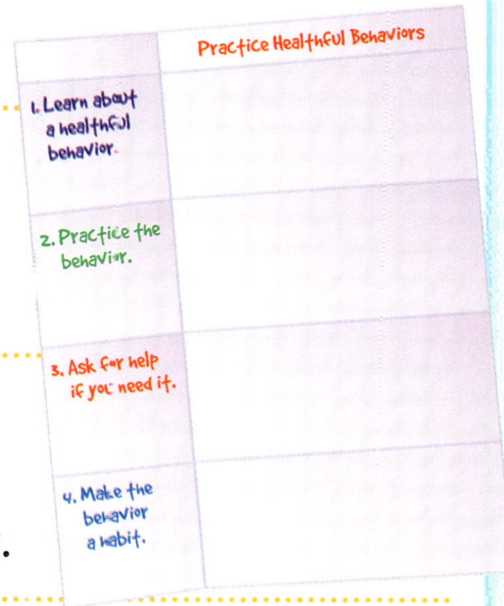

1 Learn about a healthful behavior.

You want to protect yourself from germs. You decide to learn about washing your hands.

2 Practice the behavior.

Learn when and how you should wash your hands. Practice washing your hands correctly.

3 Ask for help if you need it.

Ask your parents or guardian to remind you about washing your hands before you eat and after you use the restroom.

4 Make the behavior a habit.

Make a habit of washing your hands for a week. How did you do?

Practice This Life Skill

Write a skit about baking cookies with your parents or guardian. Explain why eating raw cookie dough is unhealthful. Act the skit out in a group in front of the class.

LESSON 3

Chronic Diseases

You will learn . . .

- about chronic diseases.
- ways to prevent chronic diseases.
- ways to show care, concern, and respect for people with special needs.

Vocabulary

- **chronic disease**, *D47*
- **allergy**, *D48*
- **asthma**, *D49*
- **diabetes**, *D50*
- **heart disease**, *D51*
- **cancer**, *D51*

Carlos plays baseball. Carlos has a chronic disease. His disease makes his leg muscles weak. He uses a wheelchair to get around. But that doesn't stop him from playing baseball.

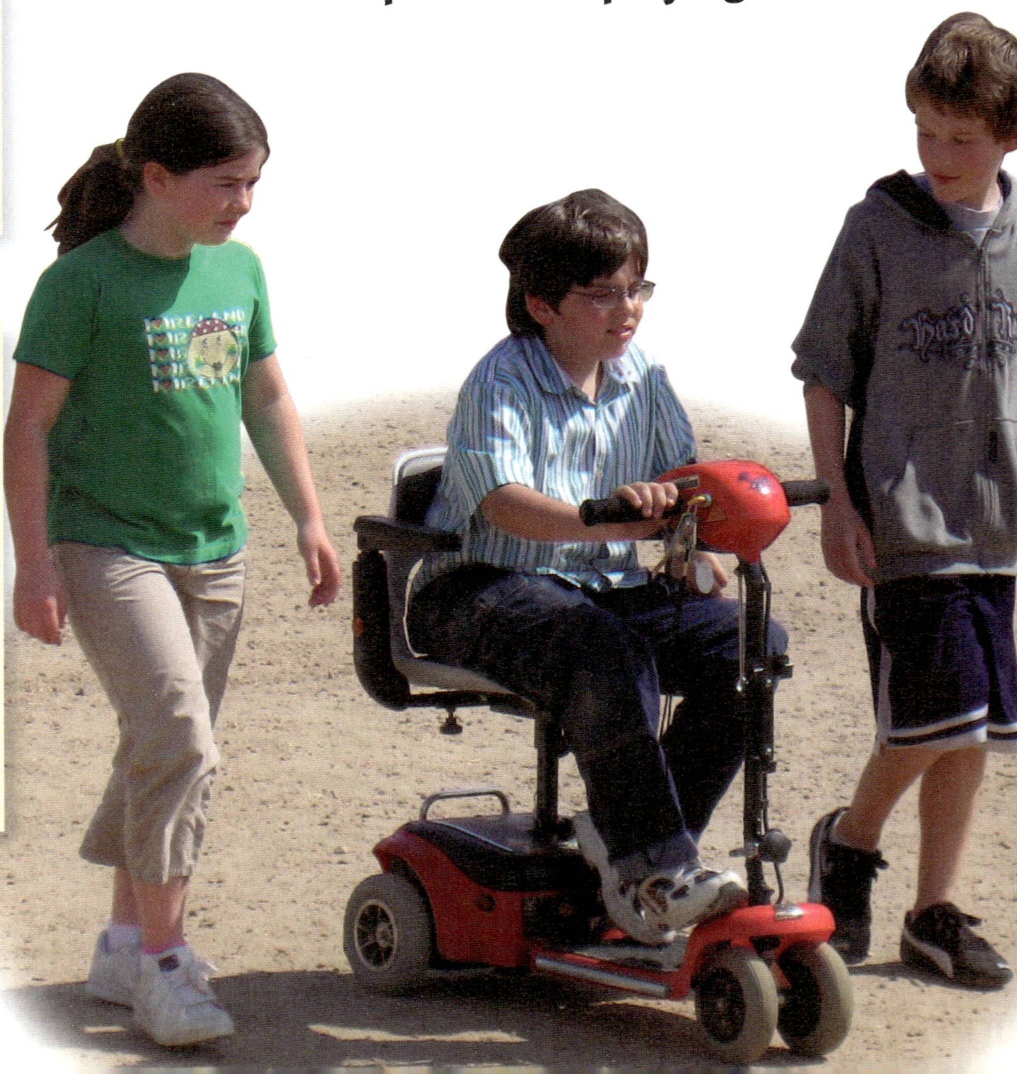

Chronic Diseases

You have learned that communicable diseases can be spread by objects, people, and animals. Some diseases cannot be spread.

A **chronic disease** is a disease that lasts for a long time or keeps coming back. Asthma and allergies are chronic diseases. Heart disease and cancer also are chronic diseases.

A child can be born with a chronic disease. The disease is passed to the child from one or both parents.

Poor health habits can lead to some chronic diseases. You can lower your risk for these diseases by practicing healthful behaviors. Some behaviors that help prevent chronic disease include:

- eating healthfully
- avoiding tobacco
- avoiding alcohol
- getting plenty of physical activity
- getting plenty of rest.

What is a chronic disease?

MAKE a Difference

Amit Bushan lives in Lubbock, Texas. One day, he had an asthma attack in a bowling alley. Breathing secondhand smoke caused the attack. Amit decided to take action against secondhand smoke. He got more than 1,000 students at his school to write letters to the city council asking for smoke-free public places. In 2001, the city council voted to make most of Lubbock's public places smoke-free!

Allergies

An **allergy** (AL•ur•jee) is the body's over reaction to a substance. Substances that cause allergies are *allergens*. Allergens are substances that cause the body to over react. Allergens include foods, animals, and plants.

Allergens You Breathe *Pollen* is a powder from plants that can get into the air. *Dust mites* are tiny insects that live in carpets, mattresses, and dust. Dust mites and pollen may cause watery eyes, sneezing, and a runny nose.

Allergens You Touch Poison ivy is a plant that causes an allergic reaction in some people. Dead skin from animal fur or bird feathers is called *dander*. Dander may cause watery eyes or a runny nose.

Allergens You Eat Some people have strong reactions to some foods, such as eggs, seafood, milk, and peanuts. These reactions include a rash, a stomachache, diarrhea, and difficulty breathing.

Preventing and Treating Allergies

The best way to prevent allergies is to avoid allergens. Doctors called allergists treat people with allergies. They may use medicines or give shots to treat allergies.

ACTIVITY

Consumer Wise

Investigate Food Labels

Reading food labels is very important for people with food allergies. Look at the labels on ten foods in your home. Make a list of foods you find that have one of the ingredients shown in the photos. What foods did you include on your list?

Eggs

Seafood

Milk

Wheat

▲ These foods are common allergens.

Asthma

Asthma (AZ•muh) is a condition in which the air passages in the lungs become narrow. During an asthma attack, the person may cough, gasp for air, wheeze, or make a sighing sound.

A person with asthma has an asthma attack only when near certain *triggers.* Some common asthma triggers are:

- allergens
- very cold environments
- strenuous exercise

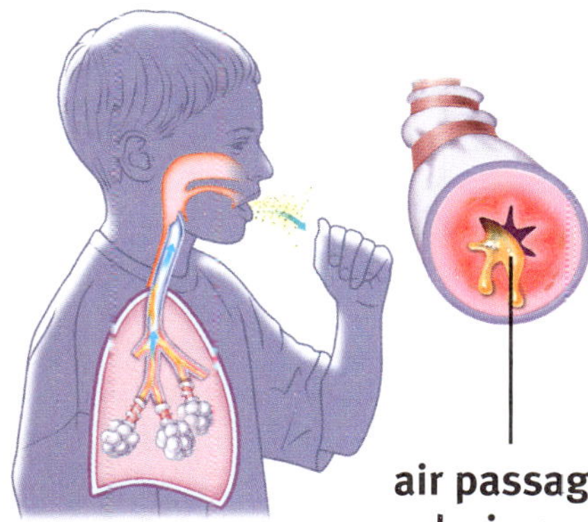

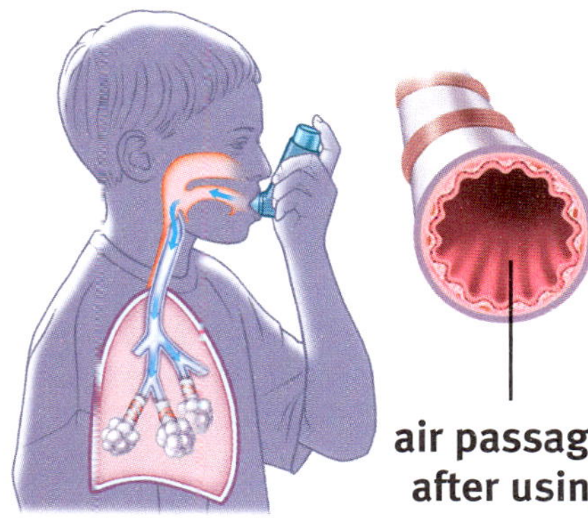

Preventing and Treating Asthma Attacks

The best way to prevent asthma is to stay away from triggers and people with colds or flu.

People with asthma may use an inhaler. An inhaler sprays medicine into the air passages to make them wider. This helps people breathe more easily.

Sometimes an inhaler does not ease an asthma attack. When this happens, medical treatment is needed.

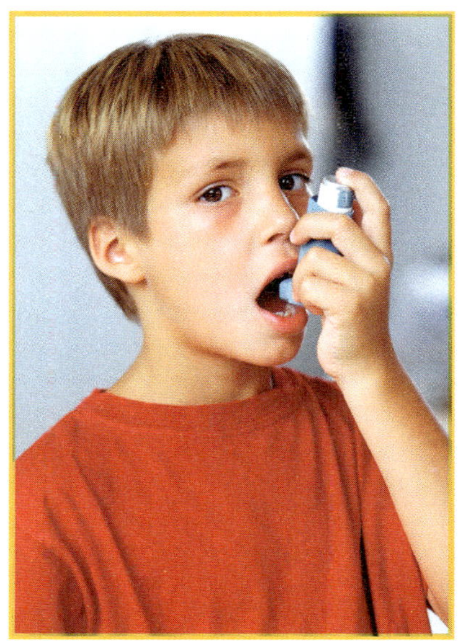

▲ The medicine in an inhaler can help widen air passages.

What is the best way to prevent an asthma attack?

Diabetes

ACTIVITY

Health Online

Types of Diabetes

Research and report on one type of diabetes. Use the e-Journal writing tool to write your report.

You get energy from food. When you eat, your digestive system changes food into sugar. Your blood carries this sugar to your cells for use as energy. Some people have diabetes. Sugar stays in their blood instead of going into the cells. **Diabetes** (digh•uh•BEE•tis) is a chronic disease in which there is too much sugar in the blood.

People who have this disease must follow a diet low in sugar. Some must also take medicine every day to help their body use the sugar in food.

People who have diabetes must test their blood often. A tool called a glucose monitor is used to test for sugar in the blood. A person can change his or her diet if the monitor tells that too much sugar is in the blood.

Some people develop diabetes in childhood. Others develop the disease later in life. You can lower your risk for developing diabetes. Three habits that lower your risk include:

- following a healthful diet
- getting plenty of physical activity
- staying at a healthful weight.

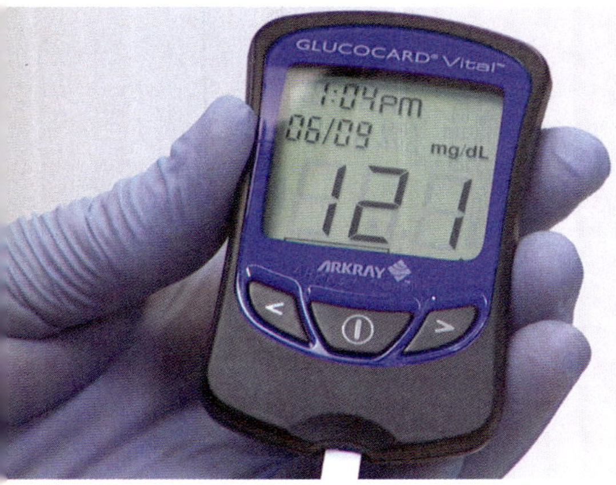

▲ **A glucose monitor is a device that lets people with diabetes check the sugar level of their blood.**

Why should a person with diabetes not eat foods that have a lot of sugar?

Heart Disease and Cancer

Heart disease is a disease of the heart or blood vessels. **Cancer** is a disease in which cells grow in ways that are not normal. You can choose habits to help prevent these chronic diseases.

Follow a Healthful Diet

You can get all the nutrients your body needs from a healthful diet. A healthful diet is rich in nutrients and low in fat. Fat can stick to your artery walls and block blood flow. If blood cannot flow through your heart, you may have a heart attack. A *heart attack* is a sudden loss of oxygen to the heart.

Fiber is the part of grains, fruits, and vegetables that helps you have a daily bowel movement. This helps prevent some cancers.

Keep Away from Tobacco

Tobacco use causes heart disease. Nicotine in tobacco makes your heart work hard by narrowing your blood vessels.

Tobacco use also causes cancer. Harmful chemicals in smoke and smokeless tobacco cause cancer of the mouth, throat, and lungs. You can prevent these cancers by keeping away from tobacco.

▼ Physical activity lowers your risk for heart disease by making your heart stronger.

People with Special Needs

CAREERS

Home Health Aide

Home health aides go to the homes of people who have medical problems. They make sure that patients get their medicines. They help patients bathe and dress. They may also help keep patients' rooms clean.

Some people have illnesses that result in a disability. A disability is not a communicable disease. A *disability* is a physical or mental condition that causes a person to have special needs. For example, a person with a spinal injury may need a wheelchair to move around. A person with hearing loss may need to use a hearing aid. A person with total hearing loss may use sign language to communicate.

You can show care and respect for a person with a disability. Treat him or her the way you would want to be treated.

How should you treat a person with a disability?

A person who is hearing impaired may use sign language.

(tl)innya/Getty Images, (br)Eclipse Studios/McGraw-Hill Education

ACTIVITY

LIFE SKILLS

CRITICAL THINKING

Make Responsible Decisions

It's a cold afternoon. LeBron asks Charles to race him to the sandbox. Charles has asthma. Running on a cold day could cause him to have an asthma attack. It would be unhealthful. Work in groups of four. Help Charles make the responsible decision.

1. **Identify your choices. Check them out with your parent or trusted adult.** Group members pick choices Charles could make. Write them at the top of a chart. Make one column for each decision.
2. **Evaluate each choice. Use the *Guidelines for Making Responsible Decisions*™.** Answer the six questions for each decision. Write the answers in the correct columns.
3. **Identify the responsible decision. Check this out with your parent or trusted adult.** The group decides which decision Charles should make. They support their decision by the answers on their chart.
4. **Evaluate your decision.** What might happen if Charles races LeBron? Decide if another decision might be more responsible for Charles to make.

LESSON REVIEW

Review Concepts

1. **Identify** two ways you can get a chronic disease.
2. **List** three sources of allergens.
3. **Name** three triggers that can cause an asthma attack.
4. **Describe** why eating healthfully helps prevent heart disease and cancer.

Critical Thinking

5. **Infer** One symptom of diabetes is feeling tired. Why might a person with diabetes feel tired?
6. LIFE SKILLS **Make Responsible Decisions** Your family is walking through the shopping mall. Your brother wants to go into a pet store. You are allergic to cats. What is the responsible decision?

CHAPTER 8 REVIEW

- **asthma,** *D49*
- **cancer,** *D51*
- **communicable disease,** *D35*
- **diabetes,** *D50*
- **fever,** *D38*
- **germ,** *D35*
- **immune system,** *D41*
- **medicine,** *D42*

Use Vocabulary

Choose the correct term from the list to complete each sentence.

1. A disease in which there is too much sugar in the blood is __?__.
2. A tiny particle that can spread disease is called a(n) __?__.
3. A disease that can be spread to people from other living things or the environment is a(n) __?__.
4. The body system that protects the body from disease is the __?__.
5. A condition in which the air passages become narrow is __?__.
6. If your body temperature is higher than normal, you have a(n) __?__.
7. A drug used to treat an illness or injury is called a(n) __?__.
8. A disease in which cells grow in ways that are not normal is called __?__.

Review Concepts

Answer each question in complete sentences.

9. Name four ways germs are spread.
10. Explain why a doctor might give you antibiotics if you have strep throat.
11. Describe the two ways your body makes antibodies.
12. Discuss what happens to a person's body when he or she has an asthma attack.
13. Name three habits that help prevent diabetes.

Reading Comprehension

Answer each question in complete sentences.

Fiber is the part of grains, fruits, and vegetables that helps you have a daily bowel movement. This helps prevent some cancers.

14. Define fiber.
15. Explain why fiber is important in a healthy diet.
16. Identify what kinds of foods provide the body with fiber.

Critical Thinking/Problem Solving

Answer each question in complete sentences.

Analyze Concepts

17. Your uncle drinks alcohol, smokes cigarettes, and never exercises. Explain why your uncle may have a greater chance of getting a chronic disease.

18. At school, your body begins to ache and you are coughing a lot. Explain how your body reacts to these types of symptoms.

19. Your grandmother has a cat. When you visit your grandmother, your eyes itch and water and your nose runs. Analyze why your body might react this way.

20. Your parents take you to the health department for a flu shot. Explain how the vaccine will help your body avoid the flu.

21. A classmate always has an assistant with her who uses sign language. Define what condition your classmate has and whether or not it is communicable.

Practice Life Skills

22. **Practice Healthful Behaviors** You have decided to practice behaviors to reduce your risk of cancer. Describe the behaviors you will make habits. Explain your choices.

23. **Make Responsible Decisions** Your friend has a cold. She wants you to play games with her. What is the responsible decision?

Read Graphics

The table describes a communicable disease common in children. Use it to answer questions 24–26.

A Communicable Disease

Disease	Symptoms	Cause	How Spread
Flu	Fever, cough, head and body aches, decreased appetite	Virus	In air (droplets released in coughs/ sneezes), hands, and touching objects that have germs on them

24. Name the disease that is caused by a virus.

25. Identify two symptoms of the disease caused by a virus.

26. Explain how the disease is spread.

Effective Communication

Write a Skit

Write a skit about two friends on opposite football teams. One friend pushes the other before he or she could catch a pass. Include how the two friends can be good sports towards one another.

Self-Directed Learning

Reading Advertisements

Go to your school library. Look in magazines for tobacco advertisements. How do the advertisements encourage smoking?

Critical Thinking and Problem Solving

Preventing Asthma Attacks

Your class is voting on a field trip. One classmate suggests the zoo. You know that your friend has asthma and allergies. Write down why the zoo would not be a healthy place for your friend to visit.

Responsible Citizenship

Make a Pamphlet

Make a pamphlet about tobacco. List the types of diseases tobacco smoke can cause and reasons you should not smoke. Put these pamphlets in the school nurse's office and in the cafeteria for students to look through.

UNIT E

Community and Environmental Health

CHAPTER 9

Consumer and Community Health

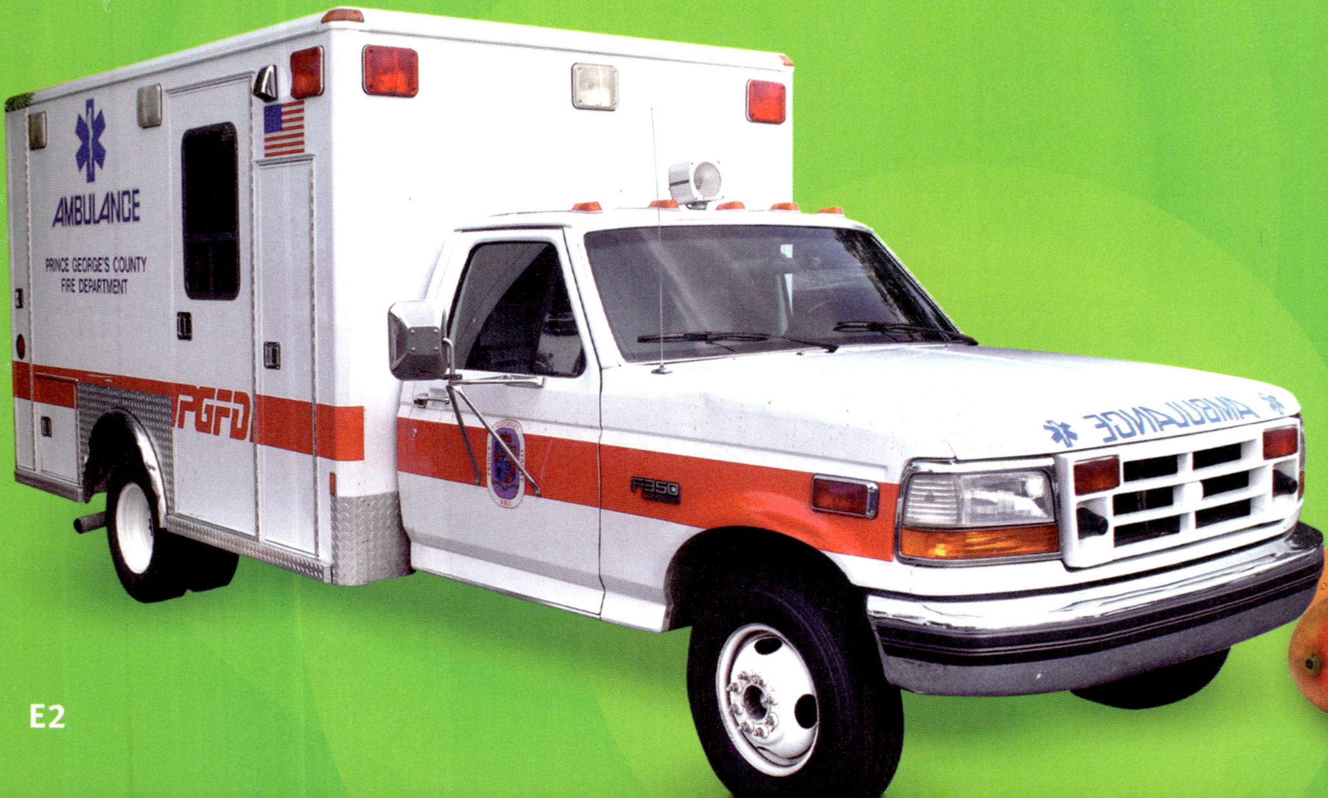

What Do You Know?

You use health products and health helpers to care for your health. Toothpaste is a health product. Your dentist is a health helper. Check off the health products and health helpers you find in the list.

- __?__ Soap
- __?__ Comb
- __?__ Firefighter
- __?__ School nurse
- __?__ Orange juice
- __?__ Police officer
- __?__ Dental floss
- __?__ Pharmacist
- __?__ Bike helmet
- __?__ Doctor

Did you check all the items on the list? You should have. Are you surprised that a bar of soap is a health product? Or that a firefighter is a health helper? Read **Consumer and Community Health** to find out more.

LESSON 1

Be a Health-Wise Consumer

You will learn . . .

- ways to be a consumer.
- ways to get health information.
- how you can check out commercials for health products.

Vocabulary

- **consumer**, *E5*
- **reliable information**, *E6*
- **media**, *E7*
- **Internet**, *E7*

You are with your parents at the supermarket. In the snack aisle, you notice a bag of low-fat tortilla chips. But wait, you also see a regular bag of tortilla chips. Which should you buy?

©Nathan Mead/age fotostock

Health Products and Services

When you shop for food and groceries with your parents or guardians at the supermarket, you are a consumer. When you choose a skateboard to buy, you are a consumer. A **consumer** buys and uses products and services. Some of the things you buy and use are health products and services.

Health Products Your parents or guardians buy health products. They may buy soap, dental floss, and fruit juice. Consumers need *health information,* or facts about health, to make responsible decisions about health products. Then a consumer can use money wisely and get the best product.

Health Services Consumers also buy health services from health helpers. People need health helpers to stay healthy. Health helpers include doctors, nurses, and dentists. Health information helps people choose the best health helpers. It helps them understand what doctors and dentists do. Knowing this information makes you a responsible consumer.

ACTIVITY

Health Online

Finding the Best Product

Receiving reliable information about a product helps consumers make wise decisions. Research and report on scientific studies done on two health products you use using the e-Journal writing tool.

▲ Your parents or guardian use health information to buy the best health products.

✔ **When do you use health information?**

Finding Health Information

ACTIVITY

On Your Own

FOR SCHOOL OR HOME

List Health Facts

Look at health product labels with an adult family member. List the facts you find. Product labels give you valuable health information. They tell you what is in the product. Labels also give directions on how to use a product properly.

Health information can come from different sources. But to be helpful, the information must be reliable. **Reliable information** is information based on scientific study. Scientists test health products and services. They decide if the information about those products and services is correct or not. Reliable information comes from sources that you can trust. You can believe what these sources tell you.

Information from Well-Known Health Groups

You can trust information that comes from well-known health groups.

- The American Heart Association and the American Cancer Society are well-known health groups.
- You can trust information that comes from the United States government. The Centers for Disease Control and Prevention (CDC) is part of the government. The CDC protects the health and safety of people. It develops preventions for all kinds of disease.

Read labels of health products with your parents or guardian to find out important health information.

Mark Dierker/McGraw-Hill Education

Best Bets for Health Information

Health groups, the U.S. government, doctors, and scientists give reliable health information in many forms. Your health textbook is another source. Your library has media, such as books, magazines, and newspapers, with health information. **Media** are sources of news and information. Radio and television are forms of media, too.

You can use the Internet to find health information. The **Internet** is a network that connects computers around the world. It has Web sites set up by groups of health helpers, such as the National Institutes for Health. Always ask permission from your parents or guardian before using the Internet. Make sure the source is reliable. Check the date of the Web site. Old health information is not always reliable.

Consumers also can take classes to learn about health. The American Red Cross offers classes in first aid. Hospitals offer health classes, too. Hospitals teach classes about nutrition, diabetes, and cancer. Some hospitals teach classes about asthma. Your teacher or school nurse may teach you about health in school.

Internet Safety Facts

Tell your parents or guardian the kind of information you want. Tell them which Web sites you plan to look at.

Never give your name, address, telephone number, or other personal information to anyone over the Internet.

Tell your parents or guardian if someone tries to find out where you live.

Tell your parents or guardian if you see strange or scary messages or pictures on the Internet.

List three sources of health information.

Analyzing Ads

ACTIVITY

Consumer Wise

Analyze Ads

Choose an ad for a health product from a newspaper, a magazine, or television. Think about how the ad tries to make you buy the product. What facts are in the ad? What facts do you think the ad leaves out? Does the ad give you enough information to make a responsible decision? Share your ad and what you think about it with the class.

Advertisements, or ads, use pictures and words to promote a product or service. Ads are meant to sell products and services. They might not give you all the facts.

You need to know why you like an ad. Some ads try to sell a product to young people. That might be the reason why you like an ad. Some ads use tricks. An athlete you like might be in an ad. You buy the product because you like the athlete. An ad might say that the product will make you popular. You buy the product because you want to be popular, too.

When you see an ad, remember that what it says may not be reliable. Think for yourself. Is the product healthful? Use reliable sources to research the product. Then decide if the product is healthful or not.

Why aren't ads sources of reliable information?

Some ads make false promises to get you to buy a product.

ACTIVITY LIFE SKILLS **CRITICAL THINKING**

Make Responsible Decisions

You saw a commercial for a glow-in-the-dark toothbrush. The toothbrush costs $20.00. Work with a group. Decide if this is a responsible purchase.

1. **Identify your choices. Check them out with your parent or trusted adult.** Make a three-column chart. List three choices you could make. Write one decision at the top of each column.
2. **Evaluate each choice. Use the *Guidelines for Making Responsible Decisions™*.** Write down your responses in the chart.
3. **Identify the responsible decision. Check this out with your parent or trusted adult.** Decide which decision is most responsible.
4. **Evaluate your decision.** Why was your choice the most responsible? Summarize your reasons.

Guidelines for Making Responsible Decisions™

- **Is it healthful?**
- **Is it safe?**
- **Does it follow rules and laws?**
- **Does it show respect for others and myself?**
- **Does it follow family guidelines?**
- **Does it show good character?**

LESSON REVIEW

Review Concepts

1. **List** three ways to be a consumer.
2. **Identify** three sources of reliable health information.
3. **Explain** how to analyze ads for health products.
4. **Name** a part of the U.S. government that gives reliable health information.

Critical Thinking

5. **Compare** Two Web sites have facts about bikes. One is a government site. The other belongs to a company that makes bikes. Tell which site has the most reliable information. Explain why.
6. LIFE SKILLS **Make Responsible Decisions** You want to buy a sports drink. An ad claims that Drink A is the most healthful. How can you make a responsible decision?

LESSON 2

Community Health Helpers

You will learn . . .

- where health helpers work.
- what health helpers do.

Vocabulary

- **health helper**, *E11*
- **community**, *E12*
- **hospital**, *E12*
- **clinic**, *E12*
- **health department**, *E12*

Your parents or guardian keep you safe and healthy at home. Your teachers help keep you healthy at school. Doctors, dentists, and health-care workers help keep you healthy, too. All of these people are health helpers.

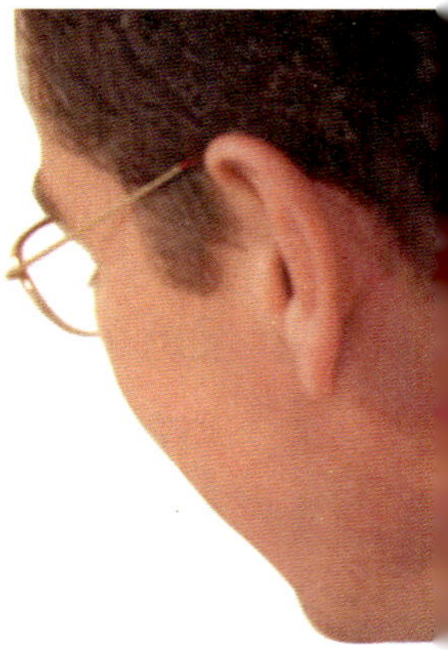

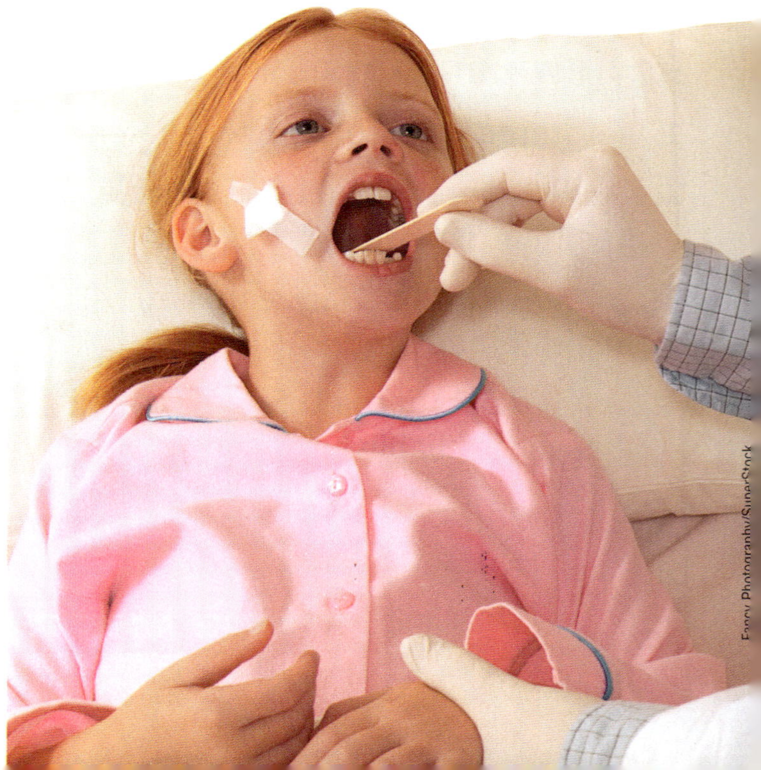

Fancy Photography/SuperStock

Home and School Health Helpers

A **health helper** is a person who helps you stay healthy. You can find health helpers at home and at school.

At School

Teachers are health helpers at school. They help you learn how to stay safe. They teach you how to stay healthy. In physical education class, they show you how to stay fit.

Your school nurse promotes health and safety. He or she gives basic first aid care for minor injuries. If there is a serious emergency the school nurse calls for help. He or she oversees you when you take your medicine at school. The nurse calls your parents or guardian if you become sick or injured.

At Home

Your parents or guardians are health helpers at home. When you are ill they give you medicine and take your temperature. They provide healthful foods for you to eat. They buy your health-care products.

Who is a health helper?

CAREERS

School Nurse

A school nurse provides basic first aid to students. He or she may give students their prescribed medicines at school. The school nurse may give students eye exams. The school nurse teaches students about health. He or she works with doctors to help students and their families manage health issues.

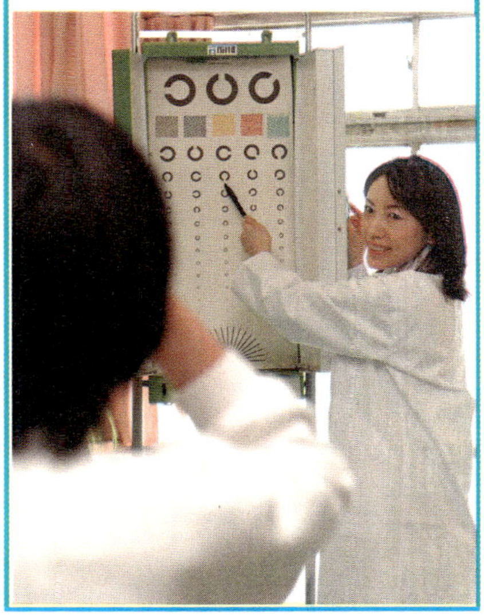

Community Health Helpers

ACTIVITY BUILD Character

Be a Community Health Helper

Citizenship You can be a community health helper. Volunteer to plant flowers at a hospital in your community. Find out if any community organizations would donate flowers. Have family members help you plant the flowers.

Your community has health helpers, too. A **community** is a place where a group of people live. Community health helpers might work in a hospital, a clinic, or a doctor's office. A **hospital** is a place where ill or injured people are treated and sometimes stay overnight. A **clinic** is a place where you can get medical care without having to stay in the hospital. People make appointments to see a doctor in his or her office.

Community health helpers also work in the health department. The **health department** oversees public health. You might get a vaccine or a flu shot at the health department.

Some Health Helpers	
Doctors treat people who are ill or injured. They work at hospitals, clinics, private offices, and health departments.	**Firefighters** put out fires and rescue people from burning buildings. They also help people in medical emergencies.
Nurses treat people who are ill or injured. They work in schools, hospitals, clinics, doctors' offices, and health departments.	**Emergency medical technicians (EMT)** take care of ill or injured people in an emergency. They may drive people to the hospital in ambulances.
Pharmacists (FAHR•muh•sist) give you the medicines your doctor tells you to take. They work in hospitals and stores called pharmacies.	**Police officers** help keep you safe by making sure people follow laws. They also help people in medical emergencies.
Dieticians plan healthful meals. They work in hospitals and schools.	**Health inspectors** check restaurants to make sure that they are clean. They work for health departments.

You can cooperate with community health helpers to stay safe. Remember to follow the laws of the community.

List three community health helpers.

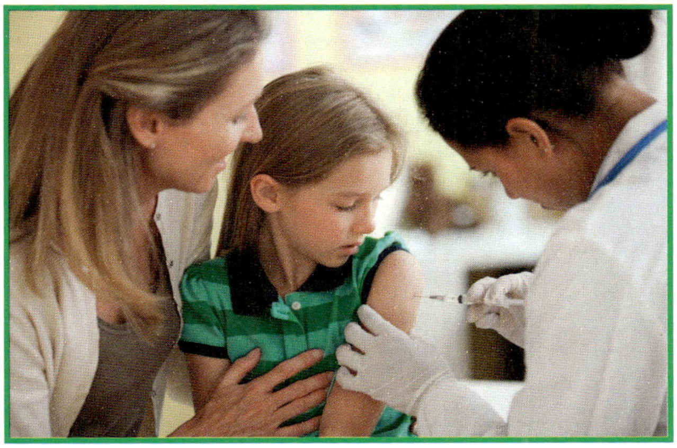

▲ **Health department workers give vaccinations to children.**

ACTIVITY LIFE SKILLS

CRITICAL THINKING

Use Communication Skills

Work in groups and role-play this scene: Lewis and Jerome are skateboarding. Lewis falls. He needs medical help. Jerome calls for help.

1. **Choose the best way to communicate.** Jerome calls 9-1-1. What other ways could he get help for Lewis?
2. **Send a clear message. Be polite.** The operator answers the phone. What should Jerome say?
3. **Listen to each other.** The operator listens to Jerome. She gives him instructions. What should Jerome do?
4. **Make sure you understand each other.** How should Jerome make sure that he understands the operator?

LESSON REVIEW

Review Concepts

1. **Identify** health helpers at your school.
2. **List** three things health helpers might do at a hospital.
3. **Explain** why it is important for a health department to communicate with schools in your community.

Critical Thinking

4. **Apply** A family member is having chest pains. Which health helpers should you call?
5. LIFE SKILLS **Access Health Facts, Products, and Services** You are at school and not feeling well. What should you do?

LESSON 3

Be an Active Community Member

You will learn . . .

- different kinds of healthful community activities.
- ways to organize your time and your money.

Vocabulary

- **organized**, *E16*

How do you choose the activities to participate in? You can choose activities that are both fun and healthful.

Healthful Activities

Taking part in community activities benefits your health. These activities can be fun, too.

Physical Activities Physical activities include skating, riding bikes, and playing baseball. These activities help keep your body healthy.

Hobbies Do you play guitar or draw pictures? Do you make objects out of recycled materials? Hobbies are healthful activities. They help keep your mind sharp.

Reading People read books, magazines, and newspapers for fun and to learn new things. Reading makes your mind strong and helps you think.

Clubs When you join a club, you can build healthful relationships. A chess club helps keep your mind active. You get a lot of physical activity in a soccer club.

What is a healthful activity?

ACTIVITY LIFE SKILLS

CRITICAL THINKING

Analyze What Influences Your Health

Play charades. Show one person trying to positively influence another. Have your classmates guess which person in the charade influenced the other's health. Use the steps below to help you analyze the charade.

1. **Identify people and things that can influence your health.**
2. **Evaluate how these people and things can affect your health.**
3. **Choose healthful influences.**
4. **Protect yourself against harmful influences.**

Have the class discuss the influences. How can healthful influences be encouraged? How can harmful influences be avoided?

▲ **Reading a book helps keep your mind active and strong.**

Being Organized

Sometimes you have a lot to do, but you run out of time. You need to be organized. When something is **organized**, it is put together in an orderly way. The tips in the chart can help you organize your time.

You can also organize your belongings. Keep your room clean. Put your things where they belong. Keep your desk clean and neat. You'll find what you need faster. You'll have more time for what you need to do.

Write About It!

Make an Organizational List Make a list of all the things you need to do tomorrow. Which things are most important to get done? Put a "1" next to those items. Put a "2" next to the things that you want to do. Put a "3" next to any items that are left. These are things that would be nice to do. Try to do all the things with "1" next to them.

Organize Your Time
• Draw a chart on a sheet of paper.
• Write down each hour of the day along the left side. Start with the time you get up.
• Include when you bathe, dress, and eat.
• Write what you will do at each time of the day. Fill in the hours you are at school.
• Write down what you do after school.
• Include time for things you need to do, like homework and chores.
• Include time for things you want to do, like visit friends, watch TV, or read a book.
• Include time to be with your family.
• Include time to be alone and rest.
• Write down the time you go to bed, too.

Manage Your Money Wisely

It is important to manage your money, too. You might get an allowance each week. You might earn money for doing chores. Save part of the money you make. Think about what you buy. Ask yourself the questions shown in the picture to the right before you buy something.

Why is it good to organize your time?

▲ Ask yourself these questions before you buy something. If you answer "no," the product is not a wise buy.

LESSON REVIEW

Review Concepts

1. **Explain** why riding a bike and playing a musical instrument are healthful activities.
2. **Discuss** how you can stay organized at school.

Critical Thinking

3. **Analyze** How can a plan help you stay organized?
4. LIFE SKILLS **Analyze What Influences Your Health** You often watch a TV show in which people only drink soft drinks. Is it a healthful influence on people who watch it?
5. LIFE SKILLS **Make Responsible Decisions** Your parents have left you with a babysitter. They told you not to watch a certain show. The babysitter doesn't know that you can't watch this show. What should you do?

Learning LIFE SKILLS

CRITICAL THINKING

Access Health Facts, Products, and Services

Problem Miranda's mother has fallen. She thinks her leg is broken.

Solution Miranda can access health facts, products, and services to get help for her mother. She can use the steps on the next page.

Foldables™ To Learn Life Skills

Learn This Life Skill

Follow these steps to access health facts, products, and services. The Foldables™ can help you.

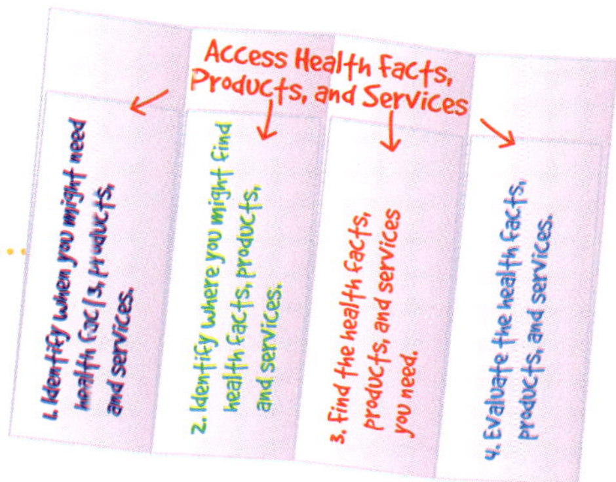

1 Identify when you might need health facts, products, and services.

Why does Miranda need help?

2 Identify where you might find health facts, products, and services.

Miranda decided to call 9-1-1. Where else might Miranda find help?

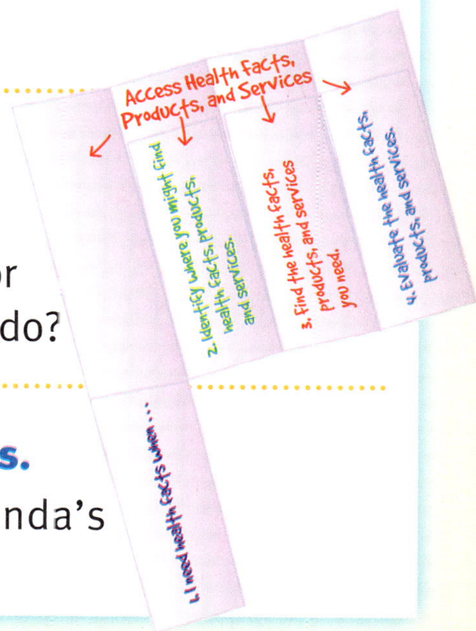

3 Find the health facts, products, and services you need.

Miranda calls 9-1-1. Miranda listens to the operator and answers her questions. What else should she do?

4 Evaluate the health facts, products, and services.

The 9-1-1 operator sends an EMT. How did Miranda's actions help improve her mother?

Practice This Life Skill

Work with a partner. Design a T-shirt about a health fact, product, or service you use. Use the steps above to guide you.

CHAPTER 9 REVIEW

Use Vocabulary

- **clinic,** *E12*
- **health department,** *E12*
- **health helper,** *E11*
- **hospital,** *E12*
- **media,** *E7*
- **organized,** *E16*
- **reliable information,** *E6*

Choose the correct term from the list to complete each sentence.

1. A(n) __?__ is a place where a person can receive medical care without having to stay in the hospital.
2. The local agency that oversees public health is the __?__.
3. Information that is based on scientific study is __?__.
4. A person who helps you stay healthy is a(n) __?__.
5. When something is __?__, it is arranged or put together in an orderly way.
6. A(n) __?__ is a place where ill or injured people may stay overnight for treatment.
7. Sources of news and information are the __?__.

Review Concepts

Answer each question in complete sentences.

8. What is a consumer?
9. When looking for health information on the Internet, what safeguards should you take?
10. List three healthful activities to keep safe and healthy.
11. Name three ways you can get organized.
12. How might community health helpers help in an emergency?

Reading Comprehension

Answer each question in complete sentences.

Community health helpers might work in a hospital, a clinic, or a doctor's office. A hospital is a place where ill or injured people sometimes stay overnight. A clinic is a place where you can get medical care without having to stay in the hospital. People make appointments to see a doctor in his or her office.

13. What is a hospital?
14. What is a clinic?
15. Where might community health helpers work?

Critical Thinking/Problem Solving

Answer each question in complete sentences.

Analyze Concepts

16. You have a rash on your arm. You aren't sure what it is. What could you do to find out what the rash is?

17. Why should you get health information from a health helper instead of relying on a commercial?

18. Your room is a mess. How can you correct this problem?

19. When might a consumer need the services of a dietician?

20. You are looking for a new dentist. How might you find one?

Practice Life Skills

21. **Access Health Facts, Products, and Services** One of your friends is injured on the school playground. Explain what you would do.

22. **Make Responsible Decisions** You want to get a new water bottle for your bicycle. Use the *Guidelines for Making Responsible Decisions*™ to tell what product to choose.

Read Graphics

23. Copy the chart below on a separate sheet of paper. Complete the chart to help you stay organized after school. Times have been added for you.

Time	Monday	Tuesday	Wednesday	Thursday	Friday
4:00					
4:30					
5:00					
5:30					
6:00					
6:30					
7:00					
7:30					
8:00					

CHAPTER 10

Environmental Health

What Do You Know?

How much do you know about the environment? Answer **true** or **false** for each statement.

__?__ Air, water, and soil are part of the environment.

__?__ The environment gives people places to enjoy.

__?__ Water and air pollution can harm the environment.

__?__ Litter can harm the environment.

Did you write **true** next to each statement? If so, you know some important facts about your environment. Read **Environmental Health** to find out more.

LESSON 1

Your Environment

What resources do you need to live? You need air to breathe. You need water to drink. You need food to eat. All of these resources come from Earth, or your environment.

You will learn . . .

- how people use natural resources.
- about the harmful effects of pollution.
- how water is kept clean.

Vocabulary

- **environment**, *E25*
- **natural resources**, *E25*
- **pollution**, *E26*
- **litter**, *E27*
- **noise**, *E27*
- **decibel**, *E27*

The Environment

What do you need to be healthy? You need fresh air, clean water, and healthful food. All of these come from your environment. The **environment** (en•VIGH•ruhn•muhnt) is everything that is around you.

Natural Resources

Natural resources are materials found in nature that are necessary or useful for life. Water, forests, metals, and oil are all natural resources.

Water Every living thing needs water. You need water to live. Plants need water to grow. Water is also used to make many products.

Forests Trees come from forests. They are used for lumber. They are also used to make paper.

Metals Iron and copper are dug from mines. Factories use metal to make products, such as refrigerators and stoves.

Oil This natural resource is a source of energy. It is pumped from under Earth's surface. Many power plants burn oil to make electricity. Oil is used to make gasoline for cars. It is also used to make plastics.

What are natural resources?

ACTIVITY

Consumer Wise

Design a Package

Think of the products you buy. Some products have a lot of packaging. Throwing away the packaging adds to trash. Choose a product that comes in a package. Would you use less packaging for the product? Draw a picture of a better package for that product.

Pollution

▲ Cars and trucks can cause air pollution.

▼ Spilled fuel can cause water pollution.

Pollution (puh•LOO•shuhn) is caused by substances in the environment that can harm health. Pollution can be in the air. It can be in the water. It can also be on or in the land.

Air Pollution

Most air pollution comes from the burning of fuels, or materials used for energy. Fuels give cars and factories energy to do their jobs. There can be air pollution inside buildings, too. Tobacco smoke causes air pollution indoors. Air pollution can cause heart and lung disease.

Water Pollution

Water pollution is the result of waste that isn't thrown away properly. *Waste* is material that is left over. Some factories might dump waste or spill fuel into the water. Rain can wash chemicals and trash into the water. People might pour poisons down the drain. Pollution can make water unsafe to drink. It can cause fish to die.

Algae are tiny plants in the ocean. *Red tides* are algae that turn the ocean a red color. Red tides make a poison that pollutes the water and air. Strong winds can carry the red tide poison in the air. This can affect the air you breathe.

Land Pollution

Land pollution is waste on or in the ground. Some communities have open dumps where trash is left on the ground. Some of the waste can leak into the ground.

Litter is trash that is thrown on the land or into the water. Litter can carry germs that make people ill.

Noise Pollution

Noise is a loud or constant sound. Noise can cause headaches. It can make you feel stress. Very loud noise can harm your hearing. A **decibel** (DEH•suh•buhl) is a measure of the loudness of sounds. The noise of many lawnmowers is about 85 decibels. Protect your hearing when noise is that loud by wearing ear plugs.

What is pollution?

ACTIVITY

Health Online

Not Just Any Dump!

Many communities have landfills for garbage. Landfills help prevent pollution. Research how a landfill is made. Use the e-Journal writing tool to write a report on your findings.

The roar of a jet can reach 140 decibels.

Clean Water

Write About It!

Write a Law Your community has laws that help stop pollution. Work with a partner. Think of a new law to help keep your community clean. Write down your law. Explain why this new law would help your community. Then make a book of your class's laws.

Turn on the faucet at home. You get clean water. Where does this water come from? Some houses have their own water wells. But most people get water from public water systems. This water comes from rivers, lakes, or wells.

Water Treatment

Water is cleaned at *water treatment plants* before people use it. At many plants, water flows into a big tank. It sits still there. Some of the dirt in the water sinks. It drops to the bottom of the tank. Cleaner water leaves the tank and flows through sand. The sand traps the rest of the dirt. It also traps some germs. Then workers add a chemical to the water. This kills germs that might still be in the water. *Fluoride* is added to water at some water treatment plants. Flouride helps protect your teeth. The water in the plant is tested often. The tests make sure that the water is healthful.

▲ **Public water systems may get water from rivers, lakes, or wells. Where does your drinking water come from?**

Where do most communities get water?

LIFE SKILLS

ACTIVITY

CRITICAL THINKING

Make Responsible Decisions

You and your friend Amy are drinking juice. Amy finishes her juice. She throws the juice box on the ground. She says, "Someone will clean it up later." Should you do the same?

1. **Identify your choices. Check them out with your parent or trusted adult.** What can you do with your juice box? List your choices.
2. **Evaluate each choice. Use the *Guidelines for Making Responsible Decisions™*.** Think about each choice. Ask yourself the six questions. Write the answers next to your list of choices.
3. **Identify the responsible decision. Check this out with your parent or trusted adult.** Review your answers. What did you decide to do with your juice box? Write down your decision.
4. **Evaluate your decision.** How is your decision responsible? How will it affect your community?

Guidelines for Making Responsible Decisions™

- **Is it healthful?**
- **Is it safe?**
- **Does it follow rules and laws?**
- **Does it show respect for me and others?**
- **Does it follow family guidelines?**
- **Does it show good character?**

LESSON REVIEW

Review Concepts

1. **Identify** a natural resource and describe how it is used.
2. **Explain** one way that air can become polluted.
3. **Describe** the way some treatment plants clean water.

Critical Thinking

4. **Analyze** Why is drinking water tested often?
5. LIFE SKILLS **Make Responsible Decisions** Your family has painted a room. You are helping to clean up. You want to get rid of some paint. You think about pouring it down the drain. Would this be a responsible decision? Explain.

LESSON 2

Protecting the Environment

You will learn . . .

- how communities stay clean and healthful.
- what you can do to stop pollution.
- ways to save natural resources.

Vocabulary

- **reduce**, *E34*
- **reuse**, *E34*
- **recycle**, *E35*

How does your community stay clean and healthful? Maybe the people in your community buy products made from recycled plastic, glass, and paper. Perhaps people volunteer to clean up your playground. They may pick up litter on the weekends.

Ariel Skelley/Getty Images

A Healthful Community

Some people in your community work to protect the environment. They work to keep the community clean and healthful.

Most communities have a sanitation (san•uh•TAY•shun) department. Sanitation workers pick up the trash. They might take the trash to a landfill. A *landfill* is a place where waste is buried. The waste is put on material that lines the landfill. The liners keep poisons in trash from leaking into the ground. Sanitation workers might also take the trash to an incinerator (in•SIN•er•ay•ter). An *incinerator* burns waste. Smoke from the incinerator is cleaned before it goes into the air.

Your community also has a health department. Workers there give people shots that help prevent diseases. They make sure restaurants are clean and safe.

How does the health department help keep the community healthful?

CAREERS

Sanitation Worker

Sanitation workers have a variety of jobs to do. They collect waste and recyclables from homes and businesses. They drive trucks along a scheduled route. They operate incinerators that burn trash. Sanitation workers work in all types of weather. They need to wear protective clothing to prevent injuries.

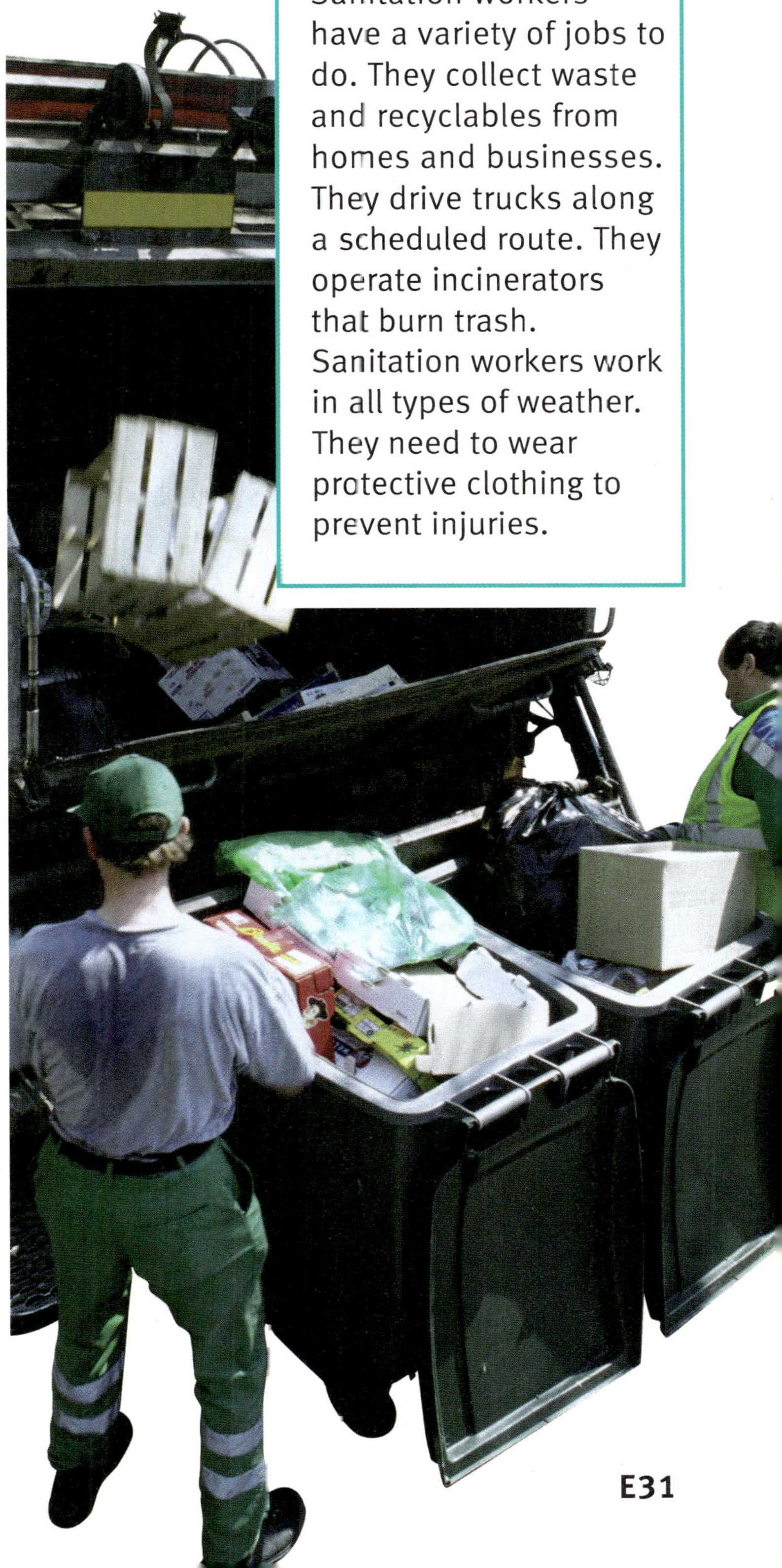

Fighting Pollution

ACTIVITY

On Your Own

FOR SCHOOL OR HOME

Take a Hike

Walking does not cause pollution. It helps keep you fit. Make a list of places you go during the week. Put a check next to the ones you walk to. Show the list to your parent or guardian.

Your community has laws to protect the environment. These laws help keep the environment clean and healthful.

Fighting Air and Water Pollution

Cars and trucks cause air pollution. People can walk places instead of taking a car. This helps keep the air clean.

Most power plants burn oil, gas, or coal to make electricity. Burning these fuels causes air pollution, too. When people use less electricity, they help keep pollution levels down. The power plants will burn less fuel and make less pollution.

Sometimes people need to get rid of paint, chemicals, or other strong cleaners. People may pour them down drains, into toilets, or on the ground. These poisons then can flow into lakes and rivers. Most communities have special collection centers to handle these wastes safely. By using your community's collection center, you fight pollution.

Pouring paint down a sink drain can cause water pollution.

Fighting Land and Noise Pollution

Litter causes land and water pollution. You can put your trash in a trash can. This helps keep your community clean.

Do you play loud music? Other people hear the noise you make. This can bother other people. Keep the sound down. This shows respect for others. This helps to keep your environment friendly. This also helps protect your own hearing. Do you listen to loud music through headphones? That can harm your hearing, too. Be sure to keep the volume at a safe level.

ACTIVITY

BUILD Character

Take Care of the Environment

Respect You show respect for other people by helping to protect the environment. Work with a partner. Plan actions you can take this week to help protect the environment. Can you turn down your radio? Can you put trash in a trash can? Each day, try one action from your plan.

How does using less electricity help protect the environment?

Putting trash in its place helps fight pollution.

Saving Natural Resources

Fighting pollution helps protect the environment. Saving natural resources does, too. You can help save natural resources. One way is to make less trash. You can reduce, reuse, and recycle.

MAKE a Difference

Reduce, Reuse, Recycle!

Students in Oregon worked hard to reduce waste. They collected 17,000 milk jugs for recycling. They used scrap paper for note paper. They returned bottles and cans to stores for recycling.

Reduce

To **reduce** is to make less of something. Another word for "reduce" is *precycle*. How could you make less trash? You could put your sandwich in a plastic container instead of plastic wrap. Remember to drink juice from a reusable bottle. Carry your lunch in a lunch box that you can wash each day.

Reuse

To **reuse** is to use items again instead of throwing them away. For example, you could use old jars and milk jugs in new ways. You could keep coins in a jar. You could plant flowers in a milk jug.

▶ **Use a lunch box instead of a brown bag to bring your lunch to school.**

Recycle

To **recycle** (ree•SIGH•kuhl) is to change trash so it can be used again. You can recycle glass jars, plastic bottles, cans, cardboard, and paper. Sanitation workers in some communities pick up materials for recycling. Other communities have recycling centers.

Some items are recycled in surprising ways. Did you know that some clothes are made of recycled plastic? Maybe the jacket you wear to school is made from recycled material.

ACTIVITY LIFE SKILLS CRITICAL THINKING

Practice Healthful Behaviors

Find a way for your class to make less trash.

1. **Learn about a healthful behavior.** Work with a group. Think about the amount of trash your class makes. What can you do to make less trash? Make a list of ideas.
2. **Practice the behavior.** Choose one or two ideas. Develop them into a plan. Tell what you will do. Tell how you will do it. Tell who will do it.
3. **Ask for help if you need it.** Show your plan to your teacher. Work out any details.
4. **Make the behavior a habit.** Make a booklet about your plan. Share it with other students at your school.

How does recycling make less trash?

LESSON REVIEW

Review Concepts

1. **Identify** workers who help keep your community healthful.
2. **Describe** one thing you can do to help stop air pollution.
3. **Define** what it means to reuse items.

Critical Thinking

4. **Infer** What would a community be like without health and sanitation workers?
5. LIFE SKILLS **Practice Healthful Behaviors** Reusing items helps keep the environment healthful. What are some ways to reuse items?

Learning LIFE SKILLS

CRITICAL THINKING

Be a Health Advocate

Problem Leaving lights on when you leave a room wastes electricity. It causes pollution. Jacob's brother often leaves the lights on. Jacob thinks this is a problem.

Solution Jacob can help his brother remember to turn off lights. He uses the four steps on the next page.

Learn This Life Skill

Follow these steps to be a health advocate. The Foldables™ can be used to help.

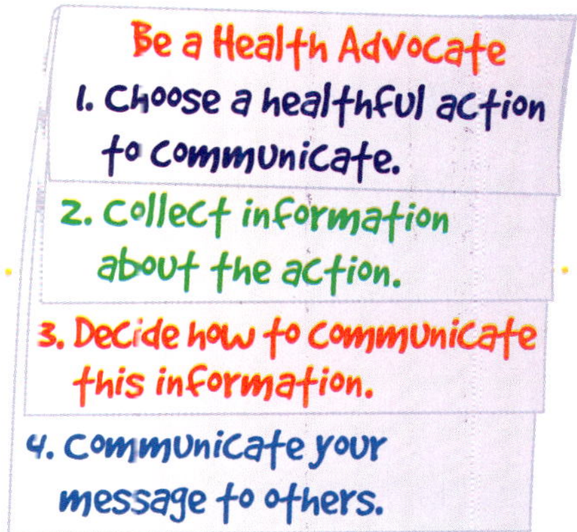

1 Choose a healthful action to communicate.

Jacob wants to help his brother save electricity.

2 Collect information about the action.

Jacob finds out why turning off lights is important to the environment. He records what he finds.

3 Decide how to communicate this information.

Jacob decides to make a poster called "save electricity."

4 Communicate your message to others.

Jacob makes his poster. He hangs it in the kitchen for everyone to see.

Practice This Life Skill

You want to explain to people why they shouldn't litter. Work with a partner. Use the four steps to be a health advocate.

CHAPTER 10 REVIEW

Use Vocabulary

decibel, *E27*
environment, *E25*
litter, *E27*
natural resources, *E25*
noise, *E27*
pollution, *E26*
recycle, *E35*
reduce, *E34*

Choose the correct term from the list to complete each sentence.

1. Trash that is thrown on land or in water is __?__.
2. To reuse trash, you must __?__.
3. Substances in the environment that can harm health are __?__.
4. A measure of the loudness of sound is a(n) __?__.
5. Everything around you is the __?__.
6. A loud or constant sound is called __?__.
7. Materials found in nature that are necessary or useful for life are __?__.
8. To __?__ is to make less of something.

Review Concepts

Answer each question in complete sentences.

9. Explain why litter can be unhealthful.
10. Name two types of pollution.
11. Why is saving natural resources important?
12. How can walking or riding a bike help prevent pollution?
13. How do sanitation workers help keep your community clean and healthful?

Reading Comprehension

Answer each question in complete sentences.

Water is cleaned at *water treatment plants* before people use it. At many plants, water flows into a big tank. It sits still there. Some of the dirt in the water sinks. It drops to the bottom of the tank. Cleaner water leaves the tank and flows through sand. The sand traps the rest of the dirt. It also traps some germs. Then workers add a chemical to the water. This kills germs that might still be in the water.

14. What is a water treatment plant?
15. Why is sand used in the plants?
16. How are the germs in the water killed?

Critical Thinking/Problem Solving

Answer each question in complete sentences.

Analyze Concepts

17. You are walking down the sidewalk. A line of firetrucks race down the street. Their sirens and horns are very loud. What should you do?

18. There is pollution in a lake. Some people fish in the lake. Why is this unhealthful?

19. A faucet in your home is dripping. You can't turn it off. What should you do?

20. A car is parked at the curb. Oil slowly drips from its engine. How could this cause water pollution?

21. You have a stack of used worksheets from school. What should you do with them?

Practice Life Skills

22. **Make Responsible Decisions** You are helping your older brother change the oil in his car. Your brother says to dump the old oil into the creek behind your house. Use the *Guidelines for Making Responsible Decisions*™ to make a responsible decision.

23. **Be a Health Advocate** Suppose you want students in your school to make less waste. Make a list of ways you can advocate for this.

Read Graphics

The table below shows how people use water. Use it to answer questions 24–25.

Water Used Each Day in Homes

Uses For Water	Number of Gallons Per Person
bathing	about 14
flushing the toilet	about 20
clothes washer	about 15
dishwasher	about 1

24. Which uses more water, washing clothes or washing dishes?

25. Based on the chart, about how much water per person is used daily?

ACTIVITIES AND PROJECTS

Effective Communication

Write a Story

Write a story about a third grader's day. Describe what he or she does during the day to help protect the environment.

Self-Directed Learning

Learn About a Career

Choose a health career that interests you. Go to the library. Read about the career. Write a report and present it to your class.

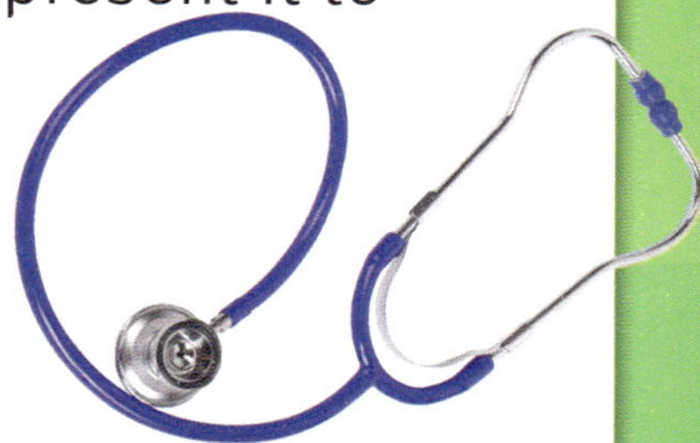

Critical Thinking and Problem Solving

Quiz a Partner

Write a set of questions about community health. Ask how health helpers work to protect the community. Ask how to prevent pollution. Trade quizzes with a classmate.

Responsible Citizenship

Make a Poster

Make a poster to promote recycling. Show how recycling helps the environment.